T0076313

WE HAVE LIFTOFF

"More than any book in recent memory, *We Have Liftoff* manages to both inform and inspire readers with a vision of space that's both inclusive and attainable. The advances in medicine it promises are especially ground-breaking. It's quite possible that future generations will look back at this book as offering *the* blueprint for what life could someday look like when we at last become a spacefaring civilization."

—Eric Hetzel, DO, District Medical Director, One Medical, Colorado Springs & Denver

"Michael Ashley and Tom Vice have created a work that is thought pro-voking, insightful, and, at times, mind bending. An excellent read for space enthusiasts and those who would rather keep their feet planted on earth's terra firma!"

—Laura R. Conover, President, Conover Consulting, Inc.

"*We Have Liftoff* is a book unlike any you will read about the coming private space race. Each chapter is written in a completely different style, demon-strating the incredible imagination of the authors in revealing what the coming years will entail. If you want to be inspired about our future, you must read this book!"

—Arie Shen, board member, technologist, serial cofounder, and operational manager

"Often we are inundated with doom-and-gloom speculations of how the future will be, with postulations of never-ending climate disasters, rogue AI, and crashing economies. In *We Have Liftoff*, we get a glimpse of the flipside of the coin. Here Michael Ashley and Tom Vice share a truly positive vision for our collective future and how human innovation and ingenuity can come together for our collective betterment."

—Julian Martinez, Executive Chef and Owner, Barbareño Restaurant

"Never before has such a strong business case been made for opening space as the final frontier. Today and tomorrow's professionals would do well to immerse themselves in this fascinating content to fully understand just how much our economy will soon transform. My message to cutting-edge industry leaders: Pick this book up now. It truly is essential reading."

—Janice L. Miller, Esq., Managing Partner, Miller Haga Law Group, LLP

"Authors Ashley and Vice make a fascinating case for why *everyone* should care about humankind's exploration of the stars. Reading this book, I was blown away by the level of detail and research the authors put into the material. A fun read, it cannot help but make you think. And more importantly—dream."

—Cindy Goss, President, Propel Business Solutions, Inc.

"If space is the final frontier, as Captain Kirk said on the starship *Enterprise*, maybe there's something to looking out there to progress our future, our medicines, and our ability to create a better society. The read made me think about all those dreams I had growing up in the '60s and what the potential is to unlocking and harnessing the energy of space. I highly recommend that you take a minute to think about what's going on 250 miles above us."

—Mark Warren, Managing Director, MFS Investment Management

"I can't think of a more ideal combination to make a 'secret sauce' of a book like *We Have Liftoff* than the amazing storyteller Michael Ashley and one of the world's most renowned space experts, Tom Vice. If you want to learn about the many cool experiential and technological advancements that are literally right around the corner (or—more precisely—right above our heads), then you must read this book. If you thought that Tang and Velcro were amazing space-related developments, you ain't seen nothing yet!"

—Mike Malatesta, the OwnerShift Coach, best-selling author
of *Owner Shift: How Getting Selfish Got Me Unstuck*

"I experienced personal liftoff reading about the future realities awaiting us in once-unreachable destinations: the moon, the stars, and galaxies of possibility beyond. The amazing worlds of Buck Rogers and Flash Gordon in my childhood comic strips and Saturday morning movies have inspired me throughout my writing career. If 'man's reach should exceed his grasp, or what's a heaven for,' what might we reach for from heaven?"

—Ron Friedman, screenwriter of more than 700 hours of top ten prime TV

"In *We Have Liftoff*, Tom Vice and Michael Ashley masterfully present an inspiring and groundbreaking panorama of the entrepreneurial opportunities lying in wait in the vast reaches of space. The authors skillfully untangle the complexities of advanced medicine, travel, and computing technology, revealing how these fields are set to undergo profound transformation in the era of space exploration. This is a road map to the future, capturing the excitement and potential of space as the new frontier for innovation and growth.

Ashley, a seasoned entrepreneur, and Vice, a well-respected expert in aerospace, harmonize their insights beautifully, delivering compelling narratives interlaced with credible data and analysis. Their contagious passion for the topic seeps through every page, making the book as inspiring as it is informative.

The beauty of *We Have Liftoff* lies in its accessibility. Whether you're a seasoned space enthusiast, a budding entrepreneur, or just curious about what the future holds, this book is a must-read. The authors present a future in which space is not merely a destination but a solution—a place where we can resolve earthly constraints and unleash the full potential of human ingenuity.

Prepare to be inspired, educated, and prompted to consider the cosmos in a new light. *We Have Liftoff* truly is a monumental leap for literature on space and entrepreneurship, igniting our imaginations and fueling our ambition for the stars. This book doesn't just explore the future, it invites us to create it."

—W. David Prescott, PG, President, Talon/LPE

"Dive into the mind-blowing world of *We Have Liftoff* by Michael Ashley and Tom Vice, a book that transforms the concept of space exploration from abstract to immediate. Melding cutting-edge science with incredibly inventive storytelling, it turns the complex topic of our future in space into something utterly captivating and digestible. Drawing on insights from the very people who are revolutionizing space travel, this book will make you feel like you're an insider on the cusp of mankind's next epoch-making leap. Get ready to see that the world of science fiction is about to materialize right in front of our eyes."

—Sameer Ahuja, President, GameChanger;
SVP, Dick's Sporting Goods

"The Orbital Age is here. *We Have Liftoff* entices me to join this exciting time in human development. The book paints a picture of reality that was once only dreamed of by science fiction writers."

—Jared Sheehan, Chief Operating Officer, QuickMD;
Director, Water Foundry

"As a kid, and now as an adult, I have been fascinated by science fiction from shows and movies like *Star Trek* and *Star Wars*. What an exciting time we live in, as I am able to see science fiction on the precipice of becoming science fact! Thank you, Michael and Tom, for a glimpse into tomorrow . . . TODAY!"

—Mark Bowling, Owner and Chief Shepherd,
BlackSheep Productions

"Truly a book to excite hearts and minds, *We Have Liftoff* will awaken you to the wonders of 21st-century space exploration. Ashley and Vice strike just the right tone between delivering essential facts and tugging on the emotions. This book could single-handedly motivate the next generation of astronauts and space workers to take to the skies."

—Joe Garner, six-time *New York Times* best-selling author

"Prepare to have your world rocked the moment you crack open *We Have Liftoff*. Written in an utterly original style, it presents a fascinating future where cutting-edge medical startups can grow organs off-world, enterprising chefs can prepare the best meal of our lives in microgravity, and Hollywood can even produce TV shows in the stars. Written for a business audience, it answers this fundamental question: Why should humanity explore the cosmos?"

—Jeff Zisselman, CEO, Sapere Vedere LLC

"What makes *We Have Liftoff* so special is how it takes the dream of space and makes it something accessible for all of humanity. Not just a select few. Authors Ashley and Vice elegantly conjure a new kind of space race, one built upon embracing what we have in common instead of what separates us. Anyone who reads it is bound to come away ever more hopeful for our future and just how far we've come."

—Bill Tedesco, CEO and Founder, DonorSearch

"For too many people, space is but an abstraction with little to no bearing upon their lives. Then, *We Have Liftoff* comes along to show modern audiences just how relevant this subject is to our lives, both now and especially in the future. Business professionals should especially take notice of Ashley and Vice's vision. Read on and you'll soon realize we're on the cusp of a civilizational sea change."

—Eric Sydell, Co-Founder and CEO, Vero AI

"As dreamers, we used to have to look to science fiction for a view of the future. Now, Michael Ashley and Tom Vice inspire a new generation of 'Dream Chasers' with a masterful blend of prescient insight and informed predictions on humanity's next big leap forward in low-Earth orbit that is both thrilling and attainable."

—Kenny Taht, CCO and Co-Founder, Plant Street Studios

WE HAVE
LIFTOFF

WE HAVE LIFTOFF

LIFTOFF

ENVISIONING YOUR PLACE
IN THE ORBITAL AGE

MICHAEL ASHLEY AND **TOM VICE**

FC

**FAST
COMPANY**
Press

Fast Company Press
New York, New York
www.fastcompanypress.com

This work is being published under the Fast Company Press imprint by an
exclusive arrangement with *Fast Company*. *Fast Company* and the *Fast Company*
logo are registered trademarks of Mansueto Ventures, LLC. The Fast Company
Press logo is a wholly owned trademark of Mansueto Ventures, LLC.

Distributed by Greenleaf Book Group

Sierra Space, LIFE, Tenacity, Ortital Age, Shooting Star, and Earth Base One are
trademarks of Sierra Space Corporation. Dream Chaser and Astro Garden are
registered trademarks of Sierra Space Corrporation.

For ordering information or special discounts for bulk purchases, please contact
Greenleaf Book Group at PO Box 91869, Austin, TX 78709, 512.891.6100.

Design and composition by Greenleaf Book Group and Kimberly Lance
Cover design by Greenleaf Book Group and Kimberly Lance
Cover image: R_Type/iStock/Getty Images Plus/Getty Images, elements of this
image furnished by NASA; and Pr3t3nd3r/iStock/Getty Images Plus/Getty
Images, elements of this image furnished by NASA

Publisher's Cataloging-in-Publication data is available.

Print ISBN: 978-1-63908-067-0

eBook ISBN: 978-1-63908-068-7

To offset the number of trees consumed in the printing of our books,
Greenleaf donates a portion of the proceeds from each printing to the Arbor Day
Foundation. Greenleaf Book Group has replaced over 50,000 trees since 2007.

Printed in the United States of America on acid-free paper

23 24 25 26 27 28 29   10 9 8 7 6 5 4 3 2 1

First Edition

Contents

Foreword

My generation was shaped by the Space Age. I grew up in the 1960s, when humanity was taking its first tentative steps beyond the Earth's gravity. Just two men had traveled to space when I was born in 1961, Yuri Gagarin and Alan Shepard. Before I turned 12, 60 people (35 American men, 24 Russian men, and 1 Russian woman) had traveled to space and 12 men (American) had walked on the moon.

Much has been written about this early history of human spaceflight—what has become known as the classical Space Age. The Cold War shaped the value proposition for the program, forged by the two superpower's attempts to outdo each other by achieving declared feats and firsts. Once beaten to the moon, the Soviet Union focused on sending cosmonauts closer to home on Earth orbital missions. The Russians built and launched seven Salyut Space Stations over the next decade, launching 70 cosmonauts who amassed nearly 2,000 days on orbit. Having won the race to the moon, the rationale for the American-manned space program was diminished. Over that same decade, NASA launched just 12 astronauts on Earth orbital missions, occupying its Skylab Space Station for only 171 days.

Although the moon landings are typically seen as the pinnacle of human spaceflight, the rationale that followed this "race" mentality has ultimately proven to be more sustainable. President Nixon's National Space Council established the goal to reduce the cost of space transportation in 1971. NASA's attempt to fulfill this mandate by developing the reusable space shuttle didn't achieve that goal but allowed the US to again surpass the Russian program, launching 355 astronauts from 16 countries (many on multiple trips) on 135 missions. In 1993 President Clinton extended an olive branch to the former Soviet Union, offering to include its former adversary in our own International Space Station (ISS).

Since the ISS initial launch in 1998, 269 people from 21 countries (many multiple times) have participated in 69 space expeditions to the space station. Humans have now been living and working in space—continuously—for the last 25 years. The generations of humans born today are living in the Orbital Age.

My recent memoir, *Escaping Gravity: My Quest to Transform NASA and Launch a New Space Age*, focuses on how more recent space policy decisions have shaped the innovations that led to where we are today. Tom Vice and Michael Ashley's new book, *We Have Liftoff: Envisioning Your Place in the Orbital Age*, describes the vast potential of what this future can bring while giving a proper name to this new era.

The Orbital Age is about more than competing over feats and firsts—it is about utilizing the unique vantage of the environment beyond Earth's atmosphere to benefit humanity. This new era of space exploration and development is as much about the journey as the destination. As Tom and Michael outline in *We Have Liftoff*, the inventions poised to be offered by more fully exploiting this unique vantage and the microgravity environment of Earth's orbit include life-saving pharmaceutical developments, critical medical technologies, advanced materials processing, critical knowledge of Earth's environment, and the transformation of our understanding of our relationship to each other, as well as our place in the universe.

Unlike the early Space Age that was driven by superpower government organizations, the Orbital Age will need to be sustained by a multitude of entities in both the public and private sectors. For now, large government programs continue to drive progress through investment in more economical and sustainable infrastructure. As these programs bear fruit and the cost of operating in the space environment is reduced, the barriers to entry are being lowered, allowing smaller government programs and commercial activities to flourish.

Similar to how governmental programs funded early exploration efforts across the seas, over the land, and above in the sky, those investments eventually seeded productive and prosperous new activities and industries. As traveling to and from space is becoming more reliable and cost effective, a multitude of transportation vehicles, laboratories, and habitats are being designed and developed to capitalize on the unique perspective, conditions, and resources available in space—just as has happened in previous exploration efforts.

We now have the knowledge, understanding, and capability to chart a course that fully utilizes the realm of space to drive progress and manage our resources to sustain humanity. This is the message of the Orbital Age, and *We Have Liftoff* tells this story beautifully. Instead of relaying detailed explanations about the scientific and technical breakthroughs space development is making possible, the book relies on a narrative of first-person stories told by current and future space travelers and inhabitants. Readers learn about Dr. Evan Townsend, a biomedical scientist researching the effects of restricted gravity on a future space station, and Maria Espinoza, a journalist covering the advances of regenerative health care on this unique facility.

Underpinned with facts, these fictional characters bring the true potential of the Orbital Age into clearer view. From new pharmaceuticals, to stem cell generated organs, to artificial retinas, the possibilities for human health advancements from space are staggering. As multinational crews explore farther away from home, the heroes of the story describe their day-to-day experiences, immersed in a new reality of space travel. From chefs to docents to dancers, our characters describe the unique attributes of the environment, which helps the reader envision how their own talents, skills, and interests could align with and benefit from this future.

Like all living species, humans explore and exploit our environment to survive. The negative impacts to our environment from previous inventions were not purposeful, but the natural result of our pursuit of progress. That innate pursuit has now led us to escape gravity and opened an entirely new venue to be explored and utilized for the benefit of humanity. Thanks to Michael Ashley and Tom Vice, we have a peek into how our progress today can lead to greater peace and prosperity tomorrow.

—Lori Garver, August 5, 2023

CHASING THE DREAM OF SPACE

As told by Mikayla Hudson, a 2029 graduate of the Sierra Space™ Astronaut Training Academy recounting her first mission as a professional astronaut. Mikayla, Mikki to her friends, launched on a Dream Chaser® named Courage. *She spent nine months supporting space workers aboard Unity Base One.*

"10—" starts the countdown by Sierra Space Mission Control.

I'm strapped securely to my chair. A good thing too, because the shaking is more intense than any simulator I've ever practiced in. I hum a song to distract myself, a trick my grandma taught me. It's from the song "Happy" by Pharrell, a favorite track back when I was a kid dreaming of today.

Suddenly, I'm back at home a few nights before takeoff.

I just shared a final dinner with my siblings, both way younger than me. My sister Jamila, then 17, picked at the barbecued shrimp I made especially for her using Grandma's recipe. Our brother Jamar,

14, didn't have the same problem—he ate it all like it was going out of style.

Jamar finally interrupted the silence. "Mikki, what am I gonna do if I need homework help while you're zooming around in space?"

I was glad he broke the ice even if his question was misinformed. "The Wi-Fi up there is better than anywhere else. I'll help you anytime you need it."

"Why are you really going to space?" he asked, dropping his fork. "To fight off aliens?"

I laughed to show him I wasn't suddenly too serious to joke around. "It's to help scientists and doctors figure out how to make medicines we can't make here. *Someone* has to keep those brainiacs safe. A scientist is liable to open up the hatch trying to find the bathroom."

Jamila finally looked up—she wasn't enjoying our exchange. "You're always so concerned about everyone else's safety. What about yours?"

"The tech is as safe as it can be, plus there's *already* a hundred people living and working in low-Earth orbit—" I had more of a speech pre-planned for her, but the moody teen cut me off before I could deliver it.

Tears in her eyes, Jamila asked, "What if you die over some stupid job and leave us all alone. Just like Dad," she adds under her breath.

That took me by surprise.

Our father, Andre Hudson, was a safety technician for an aerospace company, of all places. He died coming home from a late-night rocket test. A drunk driver claimed his life. Our mom hadn't been in the picture for a long time, leaving one teenage Mikki to help Grandma raise two little kids.

That was eight years ago.

I took Jamila's hand in mine. "Dad died doing his best to provide for all of us. I've done my best for you ever since. Now I need you to be strong—"

"But—"

". . . Because I *also* need to do my best to help the whole world. But don't worry. We'll talk daily, whether I'm 300 miles over your head, or in Timbuktu."

"9—"

I'm suddenly back at Earth Base One, the Sierra Space launch facility. Strapped into the latest Dream Chaser, known as *Courage*, it's tight. I'm reminded of the intense psychological screenings I endured at the Human Spaceflight Center. They stuffed me in the tiniest space imaginable, then left me in pitch blackness. For hours.

If I'd been at all claustrophobic, I would've clawed the walls, screaming.

But it didn't bother me then. And now? The tight space is actually a comfort. It keeps me from thinking about the massive rocket I'm sitting on.

"8—" goes the countdown, and again I fly through time.

Now, I'm back in Dad's cramped Atlanta apartment at the age of 7. My siblings weren't even born yet. Even then, he recognized something in me.

He saw my love of outer space.

I can't pinpoint when going to space became my life's focus, but it must have happened shortly after I learned what exists beyond our atmosphere. Back then Dad used to tell his friends how I favored spaceship toys over dolls.

"Other girls dress up Barbie to impress Ken," he'd say. "My

Mikki secures tools to Barbie's waist so they won't float off during spacewalks."

My dad took great pride in my space interest. He knew it wasn't a passing fad. I can't help but smile with gratitude for everything he did for me.

It didn't matter if he was tired from work, he always made time to take me to the library or the museum so I could soak up every bit of knowledge.

"Mikki," he once told me, "I've been working on rockets my whole life, but I never thought about going to space. You just keep working hard and you'll get there. No matter what anyone tells you."

"You really think I could make it to space, Dad?"

"Not a doubt in my mind."

"7—"

Back on the Dream Chaser, I feel pain in my chest. It's not from the rocket's engine. It's from thinking about him. Ten years it's been since he died. Recalling his last birthday gift, my breath catches in my throat.

He saved up for years to send me to a week-long Astronaut Camp. Showing up, I felt like the ugly duckling. All the other kids came from wealthy families. They showed up in designer clothes and pricey sneakers. They didn't like my style, my hair, or much else about me.

One girl even asked, "You sure you're in the right place?"

I faltered then, thinking they were right—like I didn't belong there. But then camp started, and the playing field suddenly evened out. We all wore the same jumpsuits, the same shoes. Before long, I impressed the other kids with my knowledge and math/engineering skills.

What first began as a test of my courage transformed into life-long friendships. A half dozen of my fellow campers wished me luck on this very mission just before I left. All had gone in different directions in life. Yet almost 20 years later, they recalled "the determined little astronaut from Atlanta."

"6—"

In a way, that character test prepared me for what happened next. The death of my father four months later.

My grandma took me in her lap the day of the funeral. "Your daddy has gone to Heaven, but he's still supporting you as he always did. He'll be looking down on you when you finish what you started . . ."

Grandma didn't have an education or the chance to build any career. But she did have a limitless well of faith. It sustained us all. Even 11-year-old me who couldn't keep her tears from falling that terrible day.

"5—"

The high school I went to wasn't the greatest, but I studied hard. While my friends binged TikTok, I kept my head in the books. One teacher saw my potential and got me transferred to a magnet school in a better neighborhood.

Here I thrived. Before long, I was bussing over to a nearby satellite campus of the Georgia Institute of Technology for college math classes, then rushing home to help Grandma raise my sister and brother—*and then* going to work at a roller rink. It wasn't outer space, but hey, at least it had laser tag.

We had no money, but every day I'd look in my mirror, telling myself (and Dad), "I'm one day closer to space."

"4—"

My hard work paid off. I was invited to tour MIT's campus and even meet with the dean of the engineering school.

She asked point blank, "So, what do you want to study?"

I immediately replied, "I'm preparing to be an astronaut."

She raised an eyebrow. "I don't think we offer a degree program in that . . ."

Her tone made it clear she wasn't attacking me, but I quickly explained myself anyway. "*Of course.* But that's my ultimate goal. The way to reach it is to learn all I can about aerospace engineering and microgravity's effect on the body. That's why I'm here. MIT is the best place to prepare for space."

That conversation must have paid off because I learned I got a full scholarship to MIT in the fall. Grandma took special care finding me a winter coat. Atlanta winters don't hold a candle to February in Massachusetts.

Glaring at the ridiculously thick parka she picked out, I protested. But as God as my witness, that winter I wore that thing daily—with layers and layers beneath—thanking my lucky stars for someone like her in my life.

My college career centered around my one and only goal. A college athlete, I still managed to work my way into every lab and the offices of each critical professor. When retired astronaut Dr. Johnathan Lum joined the aerospace faculty, he couldn't shake me, either. After a while, he stopped trying. Maybe he recognized something of himself in my zeal for space.

"3—"

Staying on at MIT, I earned a master's degree thanks to another scholarship. I began working closely with Dr. Lum on projects related to orbital mechanics and advanced materials used in space stations.

One day I got an urgent message from him. "Mikayla, I must see you. Drop what you're doing and come right now."

Anxiously entering Dr. Lum's office, I saw several others in the cramped room. One I recognized as another retired NASA astronaut.

"These people are from Sierra Space," Dr. Lum began. "They're looking for tomorrow's astronauts. They asked to meet my best and brightest prospect."

I thought my heart would stop right there. They were talking about me!

"2—"

The next two years flew by.

Taking a leave from MIT, I entered Sierra Space's Astronaut Academy. I knew of the company at the time, but I thought they mostly focused on cargo missions to existing stations. I soon learned the company had shifted to creating a new space economy of professional astronauts (what I would become) supporting space workers: private sector experts living and working in space.

After learning their mission's scope—how they planned to build many space stations in low-Earth orbit (LEO)—I signed up. *This is how I would go to space.*

Of course, Grandma was the first to learn my big news. The final people I saw as I boarded Dream Chaser *Courage* were her and my two siblings. They all waved goodbye. Happy for me, they shed no tears that day. They knew I was going where I belonged and that I would be home in nine months.

"1—"

The last second seems to hang in the air like a musical note.

I flash to Dr. Lum and the long line of teachers, advisors, and librarians who recognized my passion, who did everything they could to help me. I think of my brother and sister who'd stop their fighting long enough so I could study.

I think of my grandma who lived a tough life of loss and lack of privilege.

Most of all, I think of my father, Andre Hudson.

"WE HAVE IGNITION!"

The solid rocket motors ignite, and I'm pushed back into my seat with a powerful force I can scarcely imagine. I'm so excited, I laugh and cry at once. My tears slide backwards, welling into my ears.

Dad, I made it. I hope you're proud.

✷

GROUND CONTROL FLIES THE DREAM CHASER, but the veteran astronaut leading our flight, Captain Alex Kinley, has plenty to do. So do I, his rookie co-pilot.

Controls and information screens flank our seats. Behind us are five additional spots. Space workers fill them, including two doctors, a Nobel prizewinner, and his two younger research assistants. All look a little green around the gills. Luckily, I've prepared for weightlessness.

In fact, it's straight to business for me, exactly what I expected and trained for. My first duty is to float to the back of the ship to ensure our Shooting Star™ cargo module is properly deployed and functions normally.

The Shooting Star module sits behind the Dream Chaser. It dramatically increases the payload we can carry as well as provides a power source. Quickly, I run through a checklist I've practiced a million times before:

- Solar panels locked and ready: *check.*

- Hatch seal secure and stable: *check.*

- Energy transfer within normal range: *check.*

- Cargo stowed securely: *check.*

I radio back my results. "The Shooting Star module is within standard parameters across the board."

"Roger that," says Captain Kinley. "Why don't you take a look outside?"

In true Mikki fashion, I'd been pencils-down in my mission profile since exiting Earth's atmosphere. I finally peer out the window.

Stunned doesn't start to describe my mental state.

In *Contact*, Jodie Foster plays Dr. Ellie Arroway, an astronaut with a surprisingly similar backstory to me. She, too, had a loving father who encouraged her to go to the stars before passing away when she was young. She finally achieves her lifelong dream of journeying to space and is awed by its inexpressible beauty. "No words to describe it," she says, choking back tears. "Poetry. . . . They should have sent a poet. It's so beautiful."

Now I know just how she felt.

I'm shocked by the beauty below me. "It's like my own personal IMAX movie!" Our Nobel prizewinning scientist can only nod in mute amazement.

Nothing I might say next could ever capture the wonder I now feel.

We have some time as Mission Control lines us up with our destination. It's known to space workers by its official name of "Unity Base One," and we professional astronauts keep everything aboard functioning smoothly.

There are 24 people aboard, including five professional astronauts and 19 space workers. We'll swap places with two professionals, and our team of five space workers will replace four who are headed home, bringing the total crew to its capacity limit of 25 souls.

Captain Kinley invites everyone on the Dream Chaser to unbelt. "Time for your first true float in space."

"Can you believe it?" someone asks.

We linger for a few minutes, relishing the surreal moment.

I want to join in the fun, but Captain Kinley and I must return to work. A professional astronaut's time is scripted. You're there to do your job—not be a tourist.

Before long, our flight plan is ready. My role is to monitor multiple flight systems while also keeping an eye on our cargo module. My experience watching two unruly kids—while completing physics homework—gives me an edge.

It's soon time to dock at Unity Base One.

I'm nervous. A successful docking is like a ballet occurring 250 miles above Earth with objects moving *5 miles a second*.

After a few moments where I admittedly white-knuckle my chair, we connect with Unity Base One, issuing a satisfying thud. The rest of the procedure goes exactly according to script. In short order, we bring over our passengers and meet our new stationmates.

You've never seen someone so happy to meet a new person as when you first float aboard a space station. The longer someone has been in LEO, the more ecstatic they are to see a new face.

After the initial flurry of greetings and unloading, Captain Kinley calls me over. "This is your first time in space. Take a few minutes to yourself and enjoy it. We were all rookies once. The crew'll cover anything we need from you."

Although I love to be in the thick of the action, I do as he says. I space swim to my assigned cabin where my personal bag is strapped to a wall. Looking down on our home planet, the view is mesmerizing.

Conveniently, we are flying over North America.

I try to pinpoint my hometown, Atlanta. I don't quite understand what I'm seeing until it hits me: I am witnessing a thunderstorm from hundreds of miles above the clouds. If the Braves are playing, it'll be a rainout. Tough luck for the fans, but the storm turns my million-dollar view into something more, maybe a *billion-dollar view?*

Oh, let's be honest, the view of home from space is priceless. It'll never get old no matter how many trips I take.

After a short break, I go to work. I know just what I must do on what is officially known as an "Ingress/Egress Operation" or IEO, or what we professional astronauts call "docking day." It concerns ensuring the right supplies and materials come off the Dream Chaser and the right people and products are loaded for return. (Messing up baggage on a spaceflight is far worse than messing up the baggage on a regular airplane.)

It can result in the difference between making it home safe—or not at all.

I go about my duties feeling microgravity's profound impact, something you can prepare for but never train away. Lacking Earth's gravity, fluids spread out within you. This means they migrate more towards your upper body, making it feel like you've got permanent sinus congestion. I had the benefit of far more extensive training, so I let muscle memory take over and handle my work.

One of my favorite things about being up here is the tight bond amongst the professional astronauts aboard the station. Just as Captain Kinley let me have some alone time, we all coordinate our efforts for the good of the team. A real camaraderie exists. A true *esprit de corps* permeates our mission. Later, when a teammate is busy pulling an extra shift to repair a component, I make him lunch. I know when I'm busy fixing an ultrasound machine for a space worker, he'll brew me coffee in return.

Completing my final checklist for the day several hours later, I realize two things. One: I'm starving. Two: I'm covered in grime and feel gross. In microgravity, sweat sticks to your body in nasty blobs instead of evaporating.

Once I realize my condition, I know I must get clean. Eating in space is pretty straightforward but showering? That's a whole different ballgame.

The process is like taking a sponge bath while juggling. Only you're not manipulating regular objects, you're desperately working to keep water globules and soap packets from floating away and smacking into the face of your commanding officer as they eat dinner.

Guilty as charged. . . . *Sorry, Captain Kinley!*

The water isn't warm in the tiny screened in coatroom we call a lavatory. Yet after hours of exhausting work, I feel refreshed. I'm also so beat I can hardly keep my eyes open.

Within seconds of entering my sleeping bag mounted to the bulkhead so I don't fly off while sleeping, the intercom system goes off.

"Hudson to the command deck."

I look at my watch. 2:00 a.m. station time. I'm filled with dread. *Did I mess something up? Did I fail to secure a tool? Is Captain Kinley fuming about the soap packet to the face?*

I quickly head to the command deck, the area reserved for professional astronauts and the nerve center of our space operations. Turns out there's no problem with my work. Instead, I am to take part in a line-crossing ritual, a ceremony practiced for hundreds of years aboard ocean-going vessels. It's now made the jump to space-faring ships.

On the ocean, sailors—called *Pollywogs*—who hadn't yet sailed across the equator were transformed by crewmates into *Shellbacks* as they crossed the line. In the space version, newbies like me are called up. Together with my four veteran crewmates, we all enjoy a 90-minute orbit around our planet.

A special moment, we each pick a window to stare below in silence. My mind drifts back to an image I saw in different books as I prepared for this. *Pale Blue Dot* is the name of the iconic photograph of Earth taken 15 years before I was born by NASA's *Voyager I.*

The late astronomer Carl Sagan reflected on the image with this passage:

> Look again at that dot. That's here. That's home. That's us. On it everyone you love, everyone you know, everyone you ever heard of, every human being who ever was, lived out their lives. The aggregate of our joy and suffering, thousands of confident religions, ideologies, and economic doctrines, every hunter and forager, every hero and coward, every creator and destroyer of civilization, every king and peasant, every young couple in love, every mother and father, hopeful child, inventor and explorer, every teacher of morals, every corrupt politician, every "superstar," every "supreme leader," every saint and sinner in the history of our species lived there—on a mote of dust suspended in a sunbeam.[1]

Sagan's words capture the essence of what I see before me. Earth *is* a pale blue dot swallowed up by blackness. Everything surrounding it shrinks in color. The moon is grayish. Darkness pervades the eyeline. What leaps out is that intense blue.

When I look at land masses, I don't see many colors besides brown. You learn to recognize different colors, but even forests are muted shades. The blue glow seems to overpower everything else. Of course, it's a different story at night when the planet lights up. It's almost too beautiful to behold.

Although I'm awed by bonding with my crew—truly my brothers and sisters just like my family back home—I'm exhausted. Floating back to my cabin, I can't help but hold my head a little higher. I'm a Shellback now.

The next day is no less intense.

I'd say I hit the ground running, but in LEO you never hit the ground and you don't do much running except when strapped onto a treadmill so you don't fly off. Let's just say I floated as fast as possible without losing control.

My work aboard Unity Base One differs from that of NASA astronauts. Our dual mission is to maintain safe operations of the station while enabling our space workers to efficiently complete their work.

And let me tell you: Our space workers are up to some amazing work.

One medical team is concocting special doses of cancer-killing medication that can only be formed in microgravity. A traditional capsule later landing on Earth would destroy it all, but the Dream Chaser's gentle 1.5G touchdown will prevent that from ever happening. Another medical lab on Unity Base One is dedicated to 3D-printing organs for transplant. That's what our Nobel prizewinner is working on. He told me he hated directing the work of well-meaning astronauts from Earth, so he finally blasted off to get it right.

At the other end of the station, researchers and engineers are busy developing an exotic alloy to form crystals capable of massively boosting solar panel performance. I will later learn these space workers are working with Dr. Lum, my old mentor at MIT. (He was comically shocked when I floated into view during one of his video conferences.)

If Unity Base One was a conventional building on Earth, its Astro Garden® would be at the structure's apex. Our facility is filled with a rotation of plants producing delicious fresh food. More importantly, it might hold the answer to solving hunger terrestrially. The plants carefully bred to grow well in space are also likelier to thrive in harsh Earth locations where farming is not usually possible. Perhaps the bell peppers grown on board that I love eating raw will one day feed people back on Earth.

As a professional astronaut, it's my job to ensure everyone on board can maximize their time in LEO by being as productive as possible. Along with my normal duties of managing food supplies and constantly checking and rechecking our critical systems, I often bounce from lab to lab fixing equipment and providing advice to different teams. (Sometimes space workers just need someone to talk to for a minute that isn't their coworker.) Yes, we have strong communication systems to interface with family members and even therapists back home, but the (in-person) human touch still can't be beat.

I'm a workaholic by nature, so I try to find different ways to contribute to my crew of professional astronauts and the space workers on board Unity Base One.

The orbiting station soon becomes my home.

So, I'm honestly shocked when Captain Kinley stops by one day to tell me I'll be rotating home in a week. I can't believe nine months have passed. Only seven more days to soak in the view. But on the plus side? Only that many more days until I can see my grandma and siblings in person.

A week later I'm stowing my gear and bidding a fond farewell to the crewmates staying on board. I'm happy and sad at the same time. Sad to leave, but happy to see friends and family. Besides, I know at the rate Sierra Space is expanding, I'll be headed back to LEO in a few short months.

I fly back aboard the Dream Chaser.

The gentle runway landing of a spaceplane is one of its chief advantages. It touches down so smoothly you feel like buying the Mission Control team a few bottles of champagne. As we land at Earth Base One, I have only two things on my mind. I want to see and hug everyone, and I want some delicious hot food that won't float away on me. Upon landing, our mixed crew is led to the family reception area. The company knows how eager we are to reunite with loved ones, so they speed things along.

Just as soon as I see my grandma, everything changes. I go from being a pro astronaut into an emotional girl from Atlanta.

Jamar and Jamila run to me. I scream and attempt to dash towards them, gathering all three into a big embrace. We're all crying but I don't care. *Everyone* in the place is crying, including the ground crew.

Grandma looks the same as ever. But I'm shocked at how different my brother and sister look. It's like I've been away for years instead of months. Reunited with all of them, my mind turns to the other thing I was thinking about as we landed—fresh, delicious food. My mouth waters just thinking about a burger and fries.

Before that can happen, Grandma leans into me. "Ever since you were a little girl, your dad told me you'd be an astronaut. Now you've done it. How does it feel?"

I pause for a moment, not expecting the question. "Like I achieved my goal. And I hope I made Dad proud like you said I would."

Grandma smiles. "Oh, he's proud, you can bet on that. And so am I."

This time I don't hesitate at all. "There's one more thing: I'm going back to space as soon as they'll let me. My work there isn't done. It just got started."

The woman I adore gives me a loving smile. I answered exactly how she expected me to. Six missions later, I haven't slowed down yet.

The following is an excerpt from a podcast interview of Mikayla Hudson in the year 2030.

Q: Mikayla, why should we support more trips to space?

A: What I've learned in space is what we do out in orbit can enhance life on Earth, and one of the most direct impacts is on our health. Did you know we may soon be able to grow human organs in space? Using genetic engineering and zero gravity, we could 3D print life-saving tissue. It sounds like science fiction, but I was in the lab watching this work be done. Imagine all the people who need transplants and can't get them because of a lack of available donors. Leveraging space-based technology, we will soon be able to grow or manufacture body parts, transforming the way we provide health care.

The treatment for cancer is changing now too due to our efforts. We're able to take cancer cells to space and watch how tumors grow in microgravity. On Earth, human tissues don't form three-dimensionally like they do in space. In the coming years, cancer research labs in space will observe how these cells metastasize. Based on such insights, we will soon design drugs to inhibit this toxic growth to benefit people all over Earth.

Q: So, what about beyond medicine? Why else should people care about astronauts and space workers going to space?

A: I'll tell you. In microgravity, we can make computer chips and fiber optics essentially defect-free. Fewer defects means lower costs.

Especially if we're concerned with making sure our underprivileged communities have access to computers that can help them achieve their goals the way I did, we should want more humans in space working on this technology.

Also, space exploration will soon speed up travel on Earth. If you could go back and tell somebody in 1860 that one day people would traverse the entire United States in five hours in flying machines, they wouldn't have believed you. The best they could do back then was cross it in months—with no Netflix either!

Advancements in space exploration will soon change how we voyage in ways not unlike how the plane disrupted the train. Time is our most precious commodity, so imagine flying from California to Paris in just a few hours. This is one more of the many incredible developments our private space race promises.

Q: Interesting. What's one last thing you can tell us that would make us believe in the potential of space exploration for the benefit of all?

A: So far, we humans have explored most everywhere on Earth. We've been underwater. We've been on top of mountains. We've been to the poles. We've gone to every inhospitable place there is. Now space beckons. This is the next area for expansion. The next vista. Our next frontier.

2

WELCOME TO THE ORBITAL AGE™

Imagine this.

When Susan first awakes, she's stunned by her view of Earth, a palpable awe so magnificent, it never wears off—though she's been living and working in space for months.

A moment later, she orients herself using her cabin display to determine current time aboard the Sierra Space LIFE™ habitat, current time at her company's HQ, and current time at her home. If it isn't too late for her family back in St. Louis, she can video chat with her husband and two kids as she eats breakfast aboard the first-ever private space station.

Wait a second!

We're getting ahead of ourselves. . . . To truly appreciate the seismic shift in how people will one day work—a change on par with or greater than the internet's arrival—we must step back in time.

Humans who bore witness to the advent of steam power, the mass production of steel, electricity, and the internet could not have predicted, in real time, what a profound impact those innovations would have on our civilization. Only with hindsight can we understand the spark, the key players, the milestone moments, and the breakthrough products that define ages of innovation from the Industrial Revolution to the Information Age and beyond.

However, at this moment, there is a palpable sense that we have already entered the next and most profound period of innovation in human history. Reusable rockets and innovative fuels are lowering launch costs, making trips to LEO more affordable and rapidly more frequent. Meanwhile, space technology companies are racing to build the first commercial, on-orbit destinations to make space manufacturing a reality.

The signs are all around us: Humans are now living in the Orbital Age™.

The Orbital Age is the next industrial revolution, driven by the underlying technologies privatizing space. It is an era of historical transformation marked by the transition from 60 years of human space exploration to human space *commercialization*. In the Orbital Age, civilization shall pivot from flying a handful of astronauts to a government-run space station to transporting thousands of people on a fleet of spaceplanes to a constellation of commercial space destinations where they live and work for months at a time.

But first, to understand what's coming, we must probe the past, starting with those first pioneers. Physically impressive, the original astronauts were the best of the best. Hailing from military ranks, they had flown jet aircraft and had engineering backgrounds.

Possessing brawn, they also had brains. They knew how to use slide rulers for last-minute calculations before landing. There's even an iconic scene from the film *Apollo 13* depicting the real-life mission in which commander James Lovell (Tom Hanks) must perform complex manual math equations to save their lives after their oxygen tanks explode, marooning them in space.

This was back in the 1960s, before widespread calculator usage, personal computing, and other innovations that have gone so far to open the possibility of space exploration. Even so, modern astronaut training is no walk in the park. NASA has documented in detail the many physical ordeals even astronaut *candidates* must endure.

In 2022, Douglas Hurley recalled the difficulties of training in the Neutral Buoyancy Laboratory—the world's largest swimming pool simulating microgravity's weightlessness—in an interview with *Business Insider*. Describing how each session consisted of nonstop, physically intensive work for six hours in a pressurized suit with ill-fitting gloves, he identified his first run as one of the most physically challenging experiences he endured throughout his astronaut training.[1]

Once upon a time, *everyone* bound for space had to train under such extreme conditions. They also had to possess advanced degrees in math and engineering to respond to a host of possible emergencies (like Lovell manually plotting a course for Earth using the moon's gravity).

Today, as we enter the Orbital Age, a new category of space traveler has just emerged.

MEET . . . THE SPACE WORKER

Many intelligent people believe in humanity's future in space. But they don't yet accept that they might have a future in space *themselves*. The objection from talented professionals working in diverse fields such as biotech, materials sciences, and alternative energy may be summed up this way: "How can *I* become an astronaut? I'm far too busy pushing the boundaries of my field to put my career on hold for a year or more to make the leap to space."

Their concerns make sense.

Astronaut training *is* intense. But they're misguided. Private-sector experts needn't become full-blown astronauts of the Douglas Hurley or James Lovell variety to transform life as we know it in a hugely positive way.

They need only to become *space workers.*

Traditionally, we think of everyone who journeys to space as astronauts, but this line is blurring as private citizens venture off-world. So, what's the difference between experts, scientists, and other professionals living and working as space workers in LEO and our traditional understanding of astronauts?

It largely concerns what they'll be doing up there.

In the Orbital Age, two distinct roles in space shall emerge. The first pertains to astronauts we are already familiar with. Among other things, their duties will involve keeping the systems of the Sierra Space LIFE habitat or other private space stations working at peak efficiency and managing docking operations with spaceplane supply and crew flights such as the Dream Chaser®—all the while ensuring that personnel and equipment enjoy a smooth orbit at 17,500 miles per hour.

The number of astronauts will undoubtedly surge in the Orbital Age, especially because so many crewed space stations and other vehicles operating concurrently will also grow. But it's the second group of people, the space workers, that will balloon exponentially. The term "space worker" connotes a private-sector professional who lives and works in LEO, roughly 300 miles above our surface. Humanity will truly begin harnessing microgravity to improve life on Earth and our collective future when experts can operate off-world without needing to become full-fledged astronauts.

If this idea sounds out of this world, it's helpful to recognize that this approach has *already* been applied to other dangerous fields. Bold professionals work and live at the bottom of the ocean, on ships at sea, even in Antarctica. Many are not professional sailors or cold-weather survival specialists. Rather, they are experts in their area whose work is supported by other professionals—in this case, authorities on managing extreme conditions.

Space in the Orbital Age will follow a similar trajectory.

Now that we have established how space workers are not the same as astronauts, let's return to the typical day for the latter living and working in LEO. Susan, who we met at the start of this chapter, is an oncologist. She works for a major pharmaceutical firm. Leveraging microgravity, she hopes to develop revolutionary treatments for people suffering from the deadliest cancers.

After starting her day with a breathtaking view, much of it unfolds like it might on a work assignment in a foreign country. Still, there are differences when one inhabits the unique LEO environment.

After completing her morning routine, Susan joins her coworkers and crew for mandatory group exercise. This is not about looking good. Exercising is a necessary part of living in space. Susan knows if she doesn't work out daily for a good three hours, dramatic bone loss can and will occur, a condition called *disuse osteoporosis*.

After hitting the space gym, Susan participates in a critical safety drill. This is standard protocol so every space worker can do their part should a crisis arise. Example: how to respond if space debris from an aging satellite pierces the space station. (Preparing for the unthinkable may not be common practice for most of today's office personnel, but it will be familiar ground for those who've spent time aboard a research vessel or in Antarctica.)

These activities out of the way, Susan can focus on her research. The bulk of it concerns the type of work she excels in terrestrially. Yet as a leading cancer specialist, Susan can achieve much greater progress faster than she ever dreamed of working in an Earth-based lab. Most of all, microgravity affords her novel physical conditions to better study tumors.

Importantly, Susan is not toiling in some siloed vacuum 300 miles above all she's ever known; rather, she remains in constant contact with her team below, thanks to high-bandwidth connectivity. Susan also feels confident that the results of her experiments and lab work will return to her company's facilities via a gentle 1.5G landing aboard a Dream Chaser spaceplane.

As her day draws to a close, Susan breaks bread with her fellow space workers and astronauts—the latter responsible for 24/7 upkeep of the facilities so it remains safe and functioning. Their meal is a mixture of supplies brought from home to the station, and vegetables grown aboard the LIFE habitat. The food tastes even better than it should tonight because Susan used time on her days off to raise crops in the station's garden.

MAKING HISTORY—AS WE LIVE IT

We began our discussion with an unusual choice: upending conventional understanding of who will live and work in space. This was done to not only disrupt preconceptions but also to make you ask yourself the following: *Could I do this? Could I be a space worker? Or even a space manager?* (A discussion of space CEOs and execs will probably require a follow-up book!)

Now, let's use our initial discussion as a springboard to understand the new era coming into view at warp speed. It's safe to say many of us learned in school about humanity's distant past, when venturing beyond our atmosphere was no more than a vague dream. Textbooks will tell us humans progressed through the Stone Age, the Bronze Age, and other periods marked by technological shifts revolutionizing life on this planet. (As crude as we may consider simple bronze tools, imagine how amazing they would be in the hands of someone previously only privy to stone implements.)

Humankind's various ages share one critical thing in common: They are all recognized long *after* the fact. The pivotal era we're entering now—the Orbital Age—is one that promises to transform existence as we know it.

For context, here is a timeline of groundbreaking space exploration events that brought us to this new period of innovation:

OCTOBER 4, 1957
The Russian satellite *Sputnik 1* launches, initiating a space race between the US and the former Soviet Union.

APRIL 12, 1961
Russian astronaut Yuri Gagarin becomes the first man in space, with a 108-minute flight aboard *Vostok 1*.

MAY 5, 1961
Mercury Freedom 7 takes Alan Shepard on a suborbital flight, making him the first American in space.

MAY 25, 1961
President John F. Kennedy announces an American will land on the moon and return safely before the decade closes.

FEBRUARY 20, 1962

US astronaut John Glenn becomes the first American to orbit the Earth aboard *Mercury 6*.

JUNE 16, 1963

Russian astronaut Valentina Tereshkova becomes the first female to fly to space.

OCTOBER 11, 1968

The first crewed Apollo mission, *Apollo 7*, launches for an 11-day mission. This marks the first live TV broadcast of humans in space.

JULY 20, 1969

Apollo 11 lands on the moon, fulfilling President Kennedy's promise. Neil Armstrong steps off the ladder, declaring, "That's one small step for man, one giant leap for mankind."

APRIL 13, 1970

After an explosion ruptures *Apollo 13*'s command module, the astronauts slingshot around the moon to accelerate their safe return to Earth.

DECEMBER 19, 1972

The last mission to the moon ends with *Apollo 17* returning to Earth.

MAY 14, 1973

The United States launches Skylab, its first space station.

JUNE 18, 1983

Sally Ride becomes the first American woman in space, aboard the space shuttle *Challenger*.

JANUARY 25, 1984

In his State of the Union address, President Reagan directs NASA to construct the International Space Station (ISS) within the next decade.

JANUARY 28, 1986

Challenger explodes 73 seconds after launch, tragically killing all on board.

FEBRUARY 20, 1986

Mir, the first modular space station, launches.

NOVEMBER 20, 1998

The first segment of the ISS launches: The Zarya Control Module launched aboard a Russian Proton rocket from Baikonur Cosmodrome, Kazakhstan.

NOVEMBER 2, 2000

A combined crew of Americans and Russians become the first astronauts to reside on the ISS.

FEBRUARY 1, 2003

Space shuttle *Columbia* disintegrates upon reentry to Earth's atmosphere, tragically killing the crew.

SEPTEMBER 28, 2008

Private company SpaceX pulls off a successful launch of
Falcon 1, the first privately developed liquid-fueled rocket to
enter orbit.

JULY 21, 2011

After 30 years and 135 missions carrying humans into orbit,
NASA's Space Shuttle program ends.

2011

NASA commences a Commercial Crew Program, ushering in
a new era of commercial transport to the ISS.

NOVEMBER 25, 2015

President Obama signs the US Commercial Space Launch
Competitiveness Act. This codifies the ability of American
companies to own material resources extracted from space.

JULY 13, 2017

The Luxembourg Chamber of Deputies passes a law
guaranteeing companies the right to own resources attained
in space, making it the first European country to do so.

NOVEMBER 16, 2022

Artemis 1 saw the debut flight of NASA's Space Launch System, which is designed to return humans to the Moon.

MARCH 2023

The White House releases its National Low Earth Orbit
Research and Development Strategy, outlining how the US

can realize and institutionalize the scientific, economic, dip-
lomatic, and educational benefits of LEO research platforms
for the future.

These historical space developments—and so many more we
don't have time to go into—put into motion our new reality: that
but a few hundred miles above our heads, many people shall soon
live and work in microgravity on a semipermanent basis. The sheer
speed at which the Orbital Age is progressing makes it clear developments are occurring at an absolutely dizzying clip. Just think:
the Iron Age took roughly 1,000 years to spread across the globe.
Compare this with more recent eras, like the Late Middle Ages,
and you'll recognize humanity is now revolutionizing society in the
blink of an eye, especially as compared with antiquity.

WHAT GOES INTO AN AGE

To appreciate how far we have come so fast—and what is yet to
come—it's helpful to examine the four industrial revolutions,
bringing us to the modern era and beyond. Before doing so, let
us state that technology is forever the critical factor propelling
humankind through our various stages and ages.

But have you ever stopped to wonder: *What is technology?*

We can count on some of our brightest minds to answer this
question.

*"Technology means the systematic application of scientific or
other organized knowledge to practical tasks."*
—John Kenneth Galbraith, Economist

"Properly understood, any new and better way of doing things is technology."
　　　　　　　—**Peter Thiel**, Entrepreneur and Venture Capitalist

"Technology may be defined as the application of organized knowledge to practical tasks by ordered systems of people and machines."
　　　　　　　—**Ian Barbour**, American Scholar of Science and Religion

And, of course, our personal favorite:

"Any sufficiently advanced technology is indistinguishable from magic."
　　　　　　　—**Arthur C. Clarke**, Science Writer and Futurist

The last quote is especially applicable.

It speaks to the mentality of so many of us as we contemplate just how fast our world is changing. It seems more magical than real at times. Buckminster Fuller, the late architect, systems theorist, and designer, once described how much technology has blown apart previous conceptions of what's possible and what's fantastic.

Here he is in 1975 describing our seemingly supernatural abilities to traverse our planet—achieved in less than one lifetime.

> My father was in the leather importing business in Boston, Massachusetts, in the United States, and he imported from two places: primarily Buenos Aires and India for bringing in leathers for the shoe industry, which was centered at that time in the Boston area. And his mail, or a trip that he would like to make to Argentina,

took two months each way. And his trips to India in the mail took exactly three months each way. And it seemed absolutely logical to humanity when, early in this century, Rudyard Kipling, the English poet, said "East is East, and West is West, and never the twain shall meet." It was a very, very rare matter for any human being to make such a travel as that, taking all those months. There were not many ships that could take them there. All that has changed in my lifetime, to where I'm not just one of a very few making these circuits of the Earth, but I am one of, probably, getting to be pretty close to 20 million now, who are living a life like that around our planet.[2]

Of course, the internet has been the major innovation of our era. Moreover, unlike transportation advances, it hurtles us not through physical space but through cyberspace. Much like the tremendous agricultural advances of the early 1700s enabled industrial production, including factories and the emergence of industrial cities, so too did the web set the stage for our global economy, one in which international commerce occurs 24/7 between far-flung nations across all time zones.

In a chapter titled "How the Internet Has Changed Everyday Life," published in the book *Ch@nge: 19 Key Essays on How the Internet Is Changing our Lives*, author Zaryn Dentzel describes this dizzying historical development you no doubt lived through—but may have not realized its momentous importance. That's because we are still witnessing its earth-shattering effects.

The Internet has turned our existence upside down. It has revolutionized communications, to the extent that it is now our preferred medium of everyday communication. In almost everything we do, we use the Internet. Ordering a pizza, buying a television, sharing a moment with a friend, sending a picture over instant messaging. Before the Internet, if you wanted to keep up with the news, you had to walk down to the newsstand when it opened in the morning and buy a local edition reporting what had happened the previous day. But today a click or two is enough to read your local paper and any news source from anywhere in the world, updated up to the minute.[3]

The development of the internet occurred during what's termed the Fourth Industrial Revolution. Before describing what this era entails, let's take a survey of history to appreciate just how far we've come so fast.

1765–1870: The First Industrial Revolution

Mechanization drove a pivotal shift in agriculture, enabling machine-driven labor to replace animal and human efforts. Coal extraction, combined with the steam engine, produced a powerful new form of energy. New manufacturing activities took root, including the application of steel, textiles, and tools, generating modern industrial cities.

1870–1969: The Second Industrial Revolution

Massive technological advancements in multiple industries produced new sources of energy: electricity, oil, and gas. Assembly lines enabled mass production at scale, resulting in the internal combustion engine, along with the automobile and plane. Meanwhile, both the telegraph and telephone transformed how we communicate over vast distances. Space exploration began at the tail end of this period.

1969–1993: The Third Industrial Revolution

Extreme productivity levels created unprecedented levels of comfort and convenience across the globe, not just the developed West. Nuclear energy emerged, alongside an explosion in the usage of electronics, telecommunications, fiber optics, nanotechnology, and computers. Extensive space exploration, including travel to the moon, came online. Robotics generated staggeringly high levels of automation.

1993–Present: The Fourth Industrial Revolution

The internet's invention precipitates the modern era. Supporting IT (information technology) structures produced mass robotization, resulting in a *global* supply chain. Enabled by enhanced computer processing speeds and decreased data storage costs, AI (artificial intelligence) became commonplace tech for smartphones, vehicles, offices, and homes. Remote work took off as cloud computing and 5G connectivity empowered worldwide broadband internet access and the triumph of data as the world's most valuable commodity. Space exploration shifted from a government-funded and -conducted pursuit to the private sector, at last producing our Orbital Age.

WHY SHOULD YOU CARE ABOUT THE ORBITAL AGE?

Humanity just entered another large-scale transformative period, not unlike the Information Age. Unprecedented new commercial platforms—multiple on-orbit destinations and in-space manufacturing centers—will open an era of *space for all*, with thousands living and working in LEO. Meanwhile, broad access and a sustained space presence will lead human civilization into a new era of scientific discovery.

Commercial platforms will also allow room to grow in space as research transforms into actual production. Full-on orbit manufacturing is the next giant leap for humankind. Building new, free-flying commercial destinations in LEO, independent of the International Space Station, will make this leap possible. Just a few decades ago, the ISS opened new product research. Robust new commercial LEO destinations will now open the door to the *manufacturing* of these new products.

Here is a sampling of the exciting verticals we can expect to see in the next few years.

Cancer and Other Disease Research

Most of us have had somebody in our lives touched by cancer or some other devastating disease we've been unable to cure. So far. Space promises to change all that. According to NASA, "One of the ways the microgravity environment helps with the study of cancer cells is by removing the variable of gravity from the equation. This allows scientists to examine cancer-related cells behaving similarly to how they would act inside the human body."[4]

In a report on four decades of oncological space research, *Medical Daily* further clarifies how unique microgravity conditions can offer untold pathological breakthroughs. "When cancer cells are grown in culture plates in laboratories on Earth, they grow in a 2-dimensional flat sheet because gravity pulls them down. . . . By growing cancer cells in space, researchers have observed that they naturally form a 3-dimensional structure. In this 3D structure the cancer cells look different and behave differently than when they are grown on a flat surface, more closely mimicking tumor biology in people and animals."[5]

For centuries, cancer has been a scourge upon humankind, but innovations in space could finally release us from its death grip.

Computer Server Production

As Data Center Knowledge reports, "The server farms powering the Internet are getting super-sized, as demand for cloud services prompts technology companies to build larger and larger data center campuses."[6] A facet of the Fourth Industrial Revolution we are living through, the high-speed internet access we depend on to stream TV shows, conduct Zoom meetings, and play MMOGs (Massive Multiplayer Online Games like *World of Warcraft*) requires giant computer servers to process all that data.

Whereas server farms help to distribute processing and workload across many networked servers, data centers are the physical locations housing so many servers and networking computers.

As you might expect, all that computing gives off tremendous heat. According to a 2020 *Time* report, the number-one search engine in the world, Alphabet's Google, spends gargantuan sums to cool its servers. "Google considers its water use a proprietary trade secret and bars even public officials from disclosing the company's consumption. But information has leaked out, sometimes through legal battles with local utilities and conservation groups. In 2019 alone, Google requested, or was granted, more than 2.3 billion gallons of water for data centers in three different states, according to public records posted online and legal filings."[7]

But what if you could put data centers in the stars?

After all, the temperature in empty interstellar space hovers around 3 kelvins, not much warmer than absolute zero. Enterprising businesses are already looking to the heavens to drop their computer processing costs. The year 2021 marked the first time humankind launched a data center off-world. "The HPE Spaceborne Computer-2—a set of HPE Edgeline Converged EL4000 Edge and HPE ProLiant machines, each with an Nvidia T4 GPU to support AI workloads—was sent to the International Space Station in February of 2021," according to Data Center Knowledge.[8]

We can expect even more data centers in space, especially as costs fall.

Quantum Computing

Speaking of keeping it cool, quantum computers must be kept at near absolute zero to function. In an article for *The Quantum Authority*, tech enthusiast James Wall breaks down just how cold these devices must be to work. "For the non-chemists in the room, absolute zero means 0 Kelvin. 0 Kelvin = -273.15 Celsius = -459.67 Fahrenheit! Almost -460 degrees Fahrenheit! Think about it, the coldest temperature on record was in Antarctica at -128.6 Fahrenheit. The world literally cannot be cold enough for these computers."[9]

So far, quantum computing hasn't gone mainstream. Ask the typical person on the street to explain the difference between this and traditional computing and they're likely to be stumped. But quantum computing is a big deal, another technological juggernaut likely to transform life as we know it.

IBM has a good definition of what this entails. "Quantum computing is a rapidly-emerging technology that harnesses the laws of quantum mechanics to solve problems too complex for classical computers."[10]

Going from the theoretical to the practical, here are some ways that quantum computing may be used in the coming years:

- Complex Weather Forecasting

- Automated Financial Trading

- Cryptography

- Route and Traffic Enhancement

- Supply Chain and Inventory Optimization

- Drug Interaction Predictions

- Personalized Medicine Treatments

- Espionage

Returning to Wall's insight above, it requires specialized conditions to produce the near-absolute-zero temperature that quantum computers need to function. Enterprising organizations from IBM to Dell could spend a heap of money to produce them on Earth. Or they could set up shop in space.

At least that's what QTSPACE (Quantum Technologies in Space) believes. In the last few years, the group, composed of academics and businesspeople, has produced white papers on quantum technologies for space applications.

Here's an excerpt from one:

> Quantum Communications (QC) use the transfer of quantum information between distant terminals. One of its possible uses is quantum key distribution (QKD), which counters threats by a quantum computer on widely used asymmetric encryption, leading to long-term secure communication. The quantum secure systems developed so far provide secure communication

on ground. The extension to space and air will be the necessary complement to reach different networks of distant nodes. Essential quantum secure solutions have already been envisaged and partially developed but much more is needed in Europe for more advanced, widespread applications, and research programs.[11]

We can scarcely imagine how terrestrial life shall improve should future innovators harness quantum computing's true potential—in space.

Factories of the Future

Did you know the satellite navigation you use daily to guide your car came from space exploration? How about cordless drills? Scratch-resistant lenses? It might surprise you to learn the origins of so many of our modern inventions. Another example: Today's vertical farming craze owes a debt to NASA. For years, the agency concerned itself with discovering new ways to grow food without soil as well as with limited space and water.

The result? More hydroponic indoor farms.

Now that the costs of launching materials and equipment, as well as returning them to Earth, have lowered—and promise to fall further as rockets and fuel prices decrease—we can expect growth in space manufacturing. This may even lead to new and valuable substances, according to Mike Cruise writing for *The Conversation*. "Creating a solid foam by introducing gases into a mixture of molten glass and molten metal and allowing the mixture to cool without gravity separating the components might create a structural material with the strength of steel and the corrosion resistance of glass. But a more likely product from factories in space would be the erection of large structural sections to build further factories and space stations."[12]

Beyond originating new materials or structures, here are some other possible products to come from space-based factories:

Zeolites: for better petroleum processing and to store hydrogen gases in vehicles or stationary facilities.

Superconductors: for enhanced electricity transmission and usage in quantum computers.

Silicon Carbide: for producing unique microchips capable of working at higher temperatures and that can withstand harmful radiation.

Synthetic Diamonds: for usage in jewelry, processing chips, optics, and laser devices. Produced in microgravity, these possess higher purity and can be leveraged for industrial applications.

Carbon Nanotubes: for construction of space elevators, space suits, and vehicles based on their tremendous strength. (Estimated to be 100 times stronger than steel.)

Bulk Metallic Glass: for space structures requiring incredibly robust materials. Their applications include moon or Mars habitats, shielding against micrometeorites, space debris, and LEO collisions.

Ultra-Thin Coatings: for biocompatible coatings on implanted batteries and devices as well as photovoltaic coatings.

Space Arks: for storing data long-term at an almost unimaginable scale, that is, millions of years. Possible usages include time capsules as well as storing DNA for space reproduction cells.

Space factories remain out of reach—but only for the moment. Now that the first private space staions are coming to fruition, it's not so inconceivable that we will eventually establish off-world manufacturing centers.

Tourism

In 2001, American engineer and billionaire entrepreneur Dennis Tito dropped $20 million to go up to the ISS. For his money, he spent about a week living and working alongside astronauts. Since then, a handful of other uber-wealthy folks have self-funded their own space voyages. In 2009, Japanese fashion mogul Yusaku Maezawa, his producer Yozo Hirano, and Russian cosmonaut Alexander Misurkin took off in a Russian Soyuz capsule. Another self-funded expedition, this time to the estimated tune of $80 million, the destination was also to the ISS, lasting 12 days.

Most everyone reading this book will not be able to afford a trip costing more than the Gross Domestic Product of several small nations. But the good news is, they don't have to. Prices will be coming down, so much so that CNBC published a report in 2021 by investment bank UBS that found that "in a decade, high speed travel via outer space will represent an annual market of at least $20 billion and compete with long-distance airline flights. Space tourism will be a $3 billion market by 2030, UBS estimates."[13]

Good news for intrepid tourists, especially those without millions and millions of dollars to burn. Still, for all the money they do spend, what might such tourists expect to experience?

Here are but a few perks:

- The thrill of a rocket ship ride (way better than the world's fastest roller coaster)

- A chance to experience extended weightlessness (yes, while eating, sleeping, and even using the lavatory facilities)

- Incomparable views (not just of Earth, but of the moon, the stars, and beyond)

- A shift of perspective (it's not hyperbole to suggest this experience—not unlike a religious epiphany—will alter one's perspective and/or worldview)

- Ultimate bragging rights (you're not likely to ever come across someone else with a better vacation story)

As of this writing, some of the major companies arranging and coordinating space tourism include:

- Virgin Galactic

- Blue Origin

- SpaceX

  Note: a difference exists between orbital and suborbital spaceflights.

  Orbital: Staying in orbit, essentially traversing Earth for several days at high speeds.

  Suborbital: Shorter in duration, these usually entail a blastoff and a large arc around the Earth before returning (within approximately two to three hours).

Average costs are still quite significant for now. Virgin Atlantic charges about $250,000 for a two-hour suborbital flight. At present, it's unclear if you can write off any of these expenditures as business expenses.

Entertainment

In 2022, Citigroup analysts predicted the space industry could expect $1 trillion in annual revenue by 2040. This number also reflects launch costs falling by 95 percent.

Part of this enthusiasm includes positive media depictions of space exploration. Hollywood has been helping in this regard for years. "It is fair to say that the entertainment industry is doing a great deal to spark public interest in space travel," explained *Space Coast Daily* in March 2021. "Space themes have been ubiquitous in gaming, television, and film for a number of decades now, and will continue to be a popular draw for viewers."[14]

We are all familiar with science fiction cinematic fare, such as *Star Wars, Independence Day, Star Trek*, and *Avatar*, not to mention TV shows, such as *Space Force, Battlestar Galactica*, and *Lost in Space*. There's also a host of space-themed video games, like *Kerbal, Spaceflight Simulator, Distant Worlds*, and *SpaceBourne*.

What you may be less familiar with is how Tinseltown is expanding its studio footprint spaceward. In January 2022, *Variety* announced that a new film studio, Space Entertainment Enterprise (SEE), will soon be constructed in space. "SEE has unveiled plans to build a space station module that contains a sports and entertainment arena as well as a content studio by December 2024. . . . Named SEE-1, the module is intended to host films, television, music, and sports events as well as artists, producers and creatives who want to make content in the low orbit, microgravity environment. The facilities will enable development, production, recording, broadcasting, and livestreaming of content."[15]

SEE is apparently not alone. Another film and TV studio called Space 11 Corp has plans to enter the market. No matter which studio first plants its flag in the space entertainment realm, it may come as little surprise that none other than Tom Cruise is destined to become the first space movie star. "Just when you thought Tom Cruise achieved his greatest possible cinematic triumph with the success of the *Top Gun* sequel, the blockbuster star comes up with an idea that is, quite literally, out of this world," reported Lexie Cartwright for *New York Post* in October 2022. "The 60-year-old Hollywood veteran has reportedly teamed up with *The Bourne Identity* director Doug Liman on a movie pitch that involves filming in space, which was first tabled in 2020 before the Covid-19 pandemic halted plans."[16]

As space workers outstrip the number of astronauts living and working in the stars, we can only expect more movie and more TV productions, not just set in space, but filmed and produced there.

CAN SPACE CHANGE OUR VIEW OF REALITY?

The celebrated novelist Douglas Coupland, credited for coining the term "Generation X," once wrote, "We live in an era with no historical precedents. History is no longer useful as a tool in helping us understand current changes." There is truth to this statement. We are living in times of unprecedented transformations. The shifts are occurring so quickly it sometimes feels as if the ground is being swallowed up beneath us.

Reflecting on this moment, co-author Tom Vice presents a helpful corollary:

> I think about my two children and the environment they will soon live in. In some ways, it reminds me of the era I went through, growing up in the 1960s. We were struggling to find

meaning. We were deeply engulfed in the Vietnam War. We were questioning if we should go to space when we had so many problems in our country.

You have to visualize what the 1960s were like. It was a time of great beauty, a time of great unrest, and a time of great design—some of the most incredibly designed things came out of this era. Not just cars but space and so many other things. It was a time of people trying to find a sense of love and togetherness, and simultaneously, we were being ripped apart from each other.

One thing I do expect, though, is that when people have more chances to be in space and to look down at Earth, it will give them the sense that we all belong to this thing we call the human race. That our differences aren't so different, that we're on this tiny blue dot in this vast sea of the unknown. And if we don't get our stuff together, we're going to destroy this beautiful place.

If given the chance to look down at Earth from above, it can't help but move you, to make you think differently. I'm hoping this time of great possibility, this era we call the Orbital Age, brings us back together.

For all our knowledge and expertise, we really don't know what the beginnings of the universe are. We don't know if we're alone. It's hard to think that we are. So, if you ask me how I feel about it all, I'm optimistic. I'm worried. But I'm optimistic. If we can get millions of people to feel what it's like to look down on Earth and feel part of something bigger, I think it might help.

Clearly, the most significant industrial revolution in history is underway in space. Like the railroad, internet, and so many telecommunications advances, accelerating the commercialization of space will surely benefit life on Earth in ways we can scarcely dream of today. We must therefore seize this moment for *all* of humanity. Let us toast not just tomorrow's astronauts and space workers, but the many engineers, scientists, and explorers of all types who will produce humanity's next great space breakthroughs.

Most of all, here's to making history—even as we live it.

3

MEDICINE IN THE YEAR 2088

The following is a fictional depiction of profound medical capabilities and breakthroughs to come, told firsthand from various perspectives.

DR. EVAN TOWNSEND
Researcher. Mount Sinai Hospital, Chicago, 2024

I've spent many years in the lab working on drug discovery efforts. Always, the results come back to the tools at my disposal. It's funny. Ask the layperson, "Is there an anti-gravity machine?" Most people would say, "Yes, some sort of thing like that exists."

That's not true. Yes, there are things out there capable of *affecting* gravity—mass and distance. But no machine can remove gravity. There is no such tool at our disposal.

At least not on Earth.

To understand medicine, especially medicine's future, we must first understand gravity's importance. Everything we experience on this planet is a product of gravity exerting its force upon us. It's the reason your breakfast bar doesn't float up into the air when you open its package. The sun and moon's gravitational pull create our very ocean tides.

When it comes to health, gravity affects you every second of every day. It's forever pushing your body down, affecting your spinal health, including your posture. It's also exerting constant resistance on your bones and muscles. Remove gravity and you would immediately lose muscle mass.

Over millions of years, every terrestrial species has adapted and evolved based on constant gravitational conditions we take for granted. The materials we produce, even the way we paint, take into consideration the uniform gravity we experience here on Earth at 1G.

Now, as scientists, we like to play with things. We enjoy adjusting variables. In a lab, we can change the temperature. We can shift the pressure. We can modify the atmosphere.

What can't we do? Remove gravity. (We can increase the gravity force by putting things in a centrifuge, but we can't eliminate gravity.)

So, the one big feature about going to space next year I'm excited about on the biopharma side, is that I can restrict gravity outright.

It is a totally unique environment possessing novel conditions. There's lots of things you can do to create more ideal materials in space. Especially if you're trying to fashion new biological breakthroughs. I'll give you an example. Say we're trying to create a new organ on Earth. To do so, we must stack layers of tissue. But the more layers we add, the more problematic gravity becomes.

Conversely, from the work NASA scientists have already performed in space, it's quite clear that human anatomy responds totally differently in microgravity. The same is true for animals. Even stem cells.

They all behave and interact in unique ways once we remove this key variable. In recent years we have improved how we design biological products on Earth. We do this by optimizing and strictly controlling conditions in labs. But as I said, the one factor we've never really had control over is reducing gravity.

What will it mean for science, specifically health, when scientists and doctors *can* harness microgravity conditions at scale? Space truly is the final frontier. It opens up unprecedented ways of exploration. It permits us to fashion new materials, to better understand diseases, and to develop profound technologies.

Will it allow us to 3D-print hearts? Invent wonder drugs? We don't know yet. But I'm sure eager to find out.

MARIE ESPINOZA
Health Care Journalist. *The New England Journal of Medicine*, 2029

It still gives me chills recalling the biggest story of my career. For years, I had covered the promise of regenerative medicine. I came up at a time in which stem cells were touted as a real game-changer. The problem is, scientists just couldn't generate enough of them on Earth to truly make a medical difference.

But that was before the first space workers began taking to the skies, living and working in low-Earth orbit. Like so many others, I was skeptical about space 2.0 or whatever they were calling it. I tuned out the news of private launches, writing them off as little more than vapid publicity stunts.

But something shifted a few years into what we now call the

Orbital Age. Real progress reports leaked out to us journalists. The first big arena was manufacturing.

"Space-based R&D facilities are now announcing revolutionary stem cell advancements due to microgravity" was the lead sentence to one such article I penned. Another began like this: "The novel meets the practical in the microgravity environment. Could this portend a total paradigm shift on how biopharma companies generate fresh IP? CEO Dan Spiegel of LEO Analytics thinks so. 'We may very well soon discover *how* to create new drugs, all on account of uncovering biological processes utterly unknown on Earth.'"

Then they invited me to visit the lab. It had been two years since I interviewed Spiegel. "Come see us," he DMed me one day. "In space."

"Not on your life," I wrote back. "I still get sick on cruise ships."

"It'll be worth your while," he replied. "Or I could give someone else the exclusive . . ."

Spiegel must've known that that would push any self-respecting reporter's button. Within the week, I completed civilian training. The week after that I was space-bound. Hours later, I saw for myself the demo that would shake the medical establishment to its core.

Floating in microgravity in a hygienically sealed lab, I watched as doctors and scientists reversed the decline and death of brain cells, the ravaging process so devastating to sufferers of Alzheimer's disease.

Minutes later and still floating, I began the article that would soon be shared and reshared millions of times on the planet below: "Could today's development herald the dawn of a new health age? Might brain deterioration and so many other adverse effects of aging disappear in the coming years?"

Little did I know the prescience of these words.

ANNA NILES
Mother of Three. Atlanta, 2032

My husband Ed and I were devastated to learn our son Victor had leukemia. He was only five years old. It didn't seem fair for God to take him from us. The first week after receiving the news, I could barely keep from crying every minute of the day. I stopped eating. I stopped going to work.

Ed kept our family together. For that I am eternally grateful. Somehow, he managed to send our other two boys off to school with a packed lunch every day. He also did all the shopping and cleaning.

That period is hard to relive. Even now, tears come to my eyes as I recall holding Victor in my arms for hours. Singing to him. Rubbing his back.

My little boy was diagnosed with acute myelogenous leukemia. "There's a 65 percent survival rate," the doctor told us.

Months prior, Victor complained of feeling cold. He looked paler and was definitely not our little "climbing monkey" who used to scale the kitchen counter in a single bound. He had grown lethargic, bruising easily and often. But it wasn't until the constant nose bleeds that we suspected something was very wrong.

"AML creates too many abnormal white blood cells that just don't function right," Dr. Patel, our new oncologist, told us. "Eventually, the myeloblasts can overwhelm Victor's healthy white blood cells. Because his body's immune system needs these to fight off harmful invaders like viruses and bacteria, AML can cause infections and the flu-like symptoms you've already seen."

"Okay, but what can we do about it? Now? Before it gets too late?" asked Ed, his grip so tense on the table his knuckles had gone white.

Dr. Patel told us AML treatment usually involves a two-step process. First, Victor required remission induction therapy. This included an agonizing regimen of high-dose chemotherapy and hospitalization.

"Our number-one goal is to induce remission by destroying as many cancerous cells as we can," Dr. Patel told us.

Unfortunately, this painful process destroyed much of Victor's healthy bone marrow, preventing him from using it to replenish his body. "We're going to need a donor," Dr. Patel told us afterward. "One that's very closely matched to Victor."

No one in our family met that requirement, and there were no available donors in our small rural community. Desperate, my eldest son, Tim, created an online campaign to find a donor, but we were running out of time. Every day, Victor grew weaker and sicker.

He was at death's door when Dr. Patel called me one afternoon. I was in Victor's room with him on my lap. He had long since lost all his hair, and his face looked paler. These days, he did little besides listen to me read to him. He had gone back to liking *Cars*, just like he did when he was three. I was reading to him about Lightning McQueen and his best friend Mater, the tow truck.

"There's a new experimental treatment using synthetic but healthy blood-forming cells," Dr. Patel told me breathlessly on the videocam. "Specifically created to match Victor's DNA, these could be infused into his bloodstream."

I sat straight up. "*Synthetic?*"

"Yes, but precisely generated in space's microgravity—and at scale—to save your son."

"W-what would that do?"

"Well, usually acquiring anything even close to this in the past would've cost a fortune. If it worked at all."

"Why wouldn't it work?"

"It's complicated to explain, but basically, it's difficult to do something like this because you can't expand the cells very well. You must take a lot of cells from similar patients or donors, and the process is just not very efficient. What we found is that in space, that process works better because you remove gravity. It allows cells to expand at a rate that's far better and quicker than on Earth."

"And you think this could work for Victor?"

"I can't say for certain, but I want to try. After the infusion, the synthetic bone marrow cells should travel to Victor's bone marrow to 'jump-start' the production of healthy blood cells. But there's another benefit. It's called a graft-versus-leukemia effect. Put simply, it would force your son's immune system to view any remaining leukemia cells as foreign so he can fight them off."

Ed needed even less convincing than me. Within the hour, we agreed to the procedure.

Even now as I write these words, tears pour down my face. Only they're not tears of sadness but of joy. Because of Dr. Patel and what we jokingly now call the "miracle space treatment," my little boy just celebrated his sixth birthday.

We look forward to many more birthdays together.

DR. ROLAND EVERETTE
Co-Inventor of the Artificial Retina. New California, 2046

It's always an illuminating conversation whenever you get my grandfather and me in the room. Like his dad before him, my grandfather served as a physician all his life. Of course, back in those days, things were really twisted. Can you believe we used to allow doctors—of all people—to smoke in hospitals?

My grandfather can also tell you about the bad old days when we dispensed ether to pregnant women during deliveries. Electric shock therapy was given to mentally and emotionally impaired people. Think about lobotomies. We actually used to perform brain surgeries where doctors (and I use that word loosely here) would sever the connection between the frontal lobe and the various parts of the brain for folks with depression. Barbaric! Men and women who took the Hippocratic Oath would use a brace and a bit to drill into their patients' skulls. Jesus Christ.

Don't believe me? None other than Rosemary Kennedy, sister to our 35th president, received a lobotomy. Apparently, it was ordered by her father, Joseph Kennedy. The details are ghastly. Once an affectionate, dutiful—albeit troubled—young woman, within a year from getting the secret surgery, she was left permanently incapacitated.

But so much for the dustbin of history. Thankfully, we live in different times, an era not far from the visions of *Star Trek*, with medical doctors using diagnostic devices not unlike Dr. McCoy's tricorder.

Other predictions from that TV show have come true. For one thing, I didn't serve my full residency in an Earth-bound hospital.

Instead, a big portion of my on-the-job training occurred in a private LEO medical facility. Here, I cut my teeth treating patients with the garden-variety ailments my grandfather might have encountered at a walk-in clinic: skin rashes, broken bones, aches, pains. But also, space sickness. I even delivered a child once in microgravity. A rather messy experience I'd rather not do again.

One thing separates me from my medical antecedent family members. Great physicians all of them, none possessed any kind of capitalist streak. Not me. Guess I must have different genes. Whatever the reason, it wasn't long before I changed careers, shifting from a practicing doctor to what I like to call a *spacepreneur*.

Pretty much everyone knows me as the co-creator of the artificial retina. What they don't know as much about is how I got involved in biopharma innovation. It all began as something we now call "organ-on-a-chip tech."

As many people know, one of the issues we biopharma manufacturers run into when we design new drugs is that we don't actually get to test them on a live person until clinical trials begin.

The problem is, a company can spend close to a trillion dollars developing a drug before it ever even reaches clinical trials. Then when it finally gets to a human, a company may learn the drug doesn't work so well. Or it's dangerous.

Unfortunately, the manufacturer wouldn't have any chance to know this until this point. We all know why. It smacks of Joe Kennedy playing God with his poor daughter. Ethically, we can't just experiment on humans. We must apply a strict protocol that's hell on innovators wanting to take a drug to market.

This is a challenging situation for sure. It hurts the biopharma companies and their investors that are willing to take a risk on a new drug that could financially sink them. It also harms sick people who could benefit from a new drug if only there were a way to safely test it.

What usually happens is that manufacturers are forced to find a work-around. They must reduce the actual human to "cellular models," then use animals as surrogates for people. Not only is this not so hot from an efficacy standpoint, but it also invariably produces animal cruelty.

As a vegetarian and lifelong animal lover, I was eager to discover a new way to better test medicines on people. After assembling an amazing team of other spacepreneurs, we established our vision: to get closer to human testing—while doing no harm.

We hit upon our solution through a more complex system replicating an organ. Here's how it worked. Before we came around, imagine you were designing a drug for some cardiac disease. Your system might use mouse cells to mimic a human heart. (Of course, a mouse isn't ideal in the first place. It's quite structurally different from a human. But I digress.)

Our big idea was that we could have a complex analogue replica of a human, yet not actually test new drugs on people. Or animals. Within a few years, my partners and I got to the point where we could link many such organs on a chip together. We could have a heart linked to a liver and to a brain.

Then we took it further. There will always be a certain subset of patients suffering from certain diseases. That's why we established organ-on-a-chip technology. It's specifically designed just to test for those types of patients. It's analogous to creating highly detailed human avatars on little computer chips.

The first "guinea pigs" in our trials weren't guinea pigs at all. They were space workers. For years, we've known that people who spend much time in space often experience accelerated aging, cardiac issues, bone loss, muscle wasting, and immune system dysfunction.

Our research team created these little human organ models to replicate these problems via chips. Using them, we were able to model aging, disease, and deterioration far more efficiently and far more safely.

Looking back, I think about the bad old days of drug testing with the same horror people today probably feel when they see an old movie with a doctor lighting one up in a hospital.

Even better, our organ-on-a-chip technology benefits from machine learning. The computer modeling continues to improve by *learning* from all the space workers involved in the study. Every day we increase our study size with human-relevant tissues, building upon what we already know.

Let's talk about the idea of "building upon" because it leads to what people most know me for. As we know, the eye is a truly remarkable biological structure. The human retina possesses around 100 million photoreceptor cells.

These are spaced about one three-hundredth of a millimeter apart. Their function is to detect incoming light, focused by the eye's lens. Signals from the photoreceptor cells are then sent to the brain for analysis via the optic nerve.

Of course, there's much more that goes into seeing, but this is just to hint at the complexity of vision we take for granted. Unless, of course, something happens to one of our eyes. The point is, the sophisticated structure of the eye makes it phenomenal for seeing the world. On the other hand, its vast complexity makes it extremely hard to artificially manufacture.

Or at least it was until my team took on this project. To understand how we built the first artificial retina, we must go back to "building upon," or "stacking," as I like to call it.

It's long been known that the way to generate tissues and organs is to stack cells on top of each other. Unfortunately, those troublesome cells often don't want to stay there. Especially in gravity.

This is the reason so many terrestrial biopharma companies struggled with 3D printing back on Earth. (From your heart to your lungs, organs are mobile systems, yet you need some sort of rigid structure to build tissue or even a whole organ.)

The solution comes back to another architectural concept: create a kind of *scaffold* to hold cells in place as they mature into organs. In the past, these scaffolds worked to a limited extent. However, they also introduced foreignness to living tissue. This led to the problems you find when it comes to organ donation. The host body detects alien material as a threat and goes after it, much like your immune system attacks viruses and bacteria.

In space, the stacking problem goes away. Or, to put it more accurately, it's mitigated. Without the force of gravity—which can deform a shape just as you're printing biological materials—you don't run into the stacking problem.

In fact, in LEO you can print organs that have very fine patterns to them *and* hold their shape as they go through the maturation process. This brings us back to one of the most complex bodily structures of all, the retina. As you can probably guess by now, I'm a restless person by nature. Once we figured out the stacking deal and how to create organs on a chip at scale, I told my team we must build an artificial retina. "It is the holy medical grail."

As I mentioned, on Earth it's fine to stack cells on top of each other for a couple layers. But as soon as you get to a certain density, you get defects due to gravitational force. Worse, those defects tend to propagate. As soon as you get a flaw, it just spreads out. Wrecking everything you've built.

In space, we don't have those kinds of constraints. Using proteins, we found a way to stack charged particles on top of each other. We stacked a negatively charged particle on one side with a particle containing a positive charge on the other. Working this way, we were able to stack the thinnest of particles. This couldn't happen on Earth with gravity pushing down on such a delicate structure!

Now that's a bit of a technical explanation that might go over some people's heads, so I'll explain what we did another way. There was a 26-year-old woman who lost an eye after an accident aboard a tourist space station. If this happened on Earth, she would probably never see again. Instead, with the help of our artificial eye, her vision returned in one week. That was almost a decade ago, and still she reports 20/20 eyesight.

Of everything I've ever done, that's the proudest achievement of my life.

EMILY KEPLER
Director of Revive Project. Moon Base Sigma, 2063

The first time they replaced any of my body parts, I was nine. I suffer from an anomalous coronary artery. A congenital heart defect, it first manifested when I was a baby. My mom later told me that I had trouble breathing and that my legs and ankles would swell. These problems seemed to go away until the fourth grade, when they resurfaced with a vengeance. After reporting chest pains to the pod nurse, they transported me to Rikers Ido, a brand-new facility nestled in the South Pole–Aitkin Basin crater. By the time my parents joined me, a *healthbot* was busy conducting tests on me, both an X-ray and an ECG. Afterward, we teleconferenced with Dr. Aronson, a cardiologist back on Earth, who delivered the official diagnosis.

"Emily," said Dr. Aronson, beaming in via a com-link, "just a few years ago, doctors had little recourse when a young person such as yourself suffered from heart disease. The typical treatments included medicine to aid your heart with pumping or even oxygen therapy. Of course, we could also employ surgical treatments to remedy the defect . . ."

Mom and I locked eyes. For three years she had supported me in my fierce desire to be a gymnast: flying me to practice, getting me the latest vitamin infusions. All that preparation would come to nothing, my dream disappearing, if I took Dr. Aronson up on whatever medical procedure he had in mind.

"No," I blurted out.

My mom shook her head. "But you haven't even heard Dr. Aronson out yet."

"I don't care what he says. I'm not giving up gymnastics. I'm not."

My dad approached me, taking my hand in his. "Emily, dear. I don't think you understand how serious this is. We could lose you if you don't do exactly what Dr. Aronson suggests."

By now I was crying into his shoulder and nearly inconsolable.

Dr. Aronson waited until my outburst was over to deliver his suggestion. It surprised all of us. "We can fix your heart today— and it won't slow you down a beat from being the moon's most famous gymnast."

I couldn't believe what I was hearing. "I can still be a gymnast?"

"How?" Dad cut in. "There's less than 500 people in this lunar region. How could we possibly get a transplant today?"

Dr. Aronson smiled. "I'm not talking about a transplant." He looked straight at me. "We can *replace* your heart with your own cells. After, of course, we use gene editing to remove any deleterious anomalies."

I didn't know what those last few words meant, but it didn't dampen my mood. As Dr. Aronson continued, explaining concepts like biocompatibility and mitigating the risk of implant rejection, my mind was on to my next meet. I saw myself on the balance beam, gracefully executing handsprings and back handsprings, saltos, back saltos, turns, and split jumps—

". . . There's a chance Emily's new heart won't just free her from maladies like chest pains, but that it will actually work better than the one she has now." I caught the tail end of Dr. Aronson's words.

"*Better?* Better how?" I asked.

"I'll just put it this way," he grinned with those twinkly eyes of his, "come back to me in a few days and tell me how you feel."

The operation was swift and mostly painless. I was under anesthesia the whole time, so really, the only uncomfortable part was the needle going into my arm. When I awoke, I remember feeling some heaviness in my chest, but it was mild and went away before the week was out.

But that wasn't the interesting part. At least, not for my nine-year-old self. The amazing thing was how I performed the next time I hit the balance beam. The best way to describe it is the silliest I know, but it's true. I felt like a . . . superhero. It's like I had new powers suddenly.

When I reported this back to Dr. Aronson, he was his typical smiley self. "Do go on . . ."

With Mom and Dad watching me, I tried to explain how I felt. "It's like there was this other version of me before the operation. And that was a good version. But this . . . this is a way *better* version. I feel lighter, stronger. I don't get tired so much. I feel just . . . powerful."

"What's going on, Doc?" Dad said in his jokey voice. "What'd you do to my little girl?"

The moment he asked his question, a nearby healthbot engaged. As it did, Dr. Aronson projected a hologram into midair. It looked just like the movies I liked to watch, except the image was a three-dimensional representation of my heart.

"To fully appreciate what happened to Emily, we need to know a bit about human anatomy. On a cellular level, we've long known that cardiac stem cells regenerate all the time. Obviously, when Emily was just a baby in her mother's womb, her first heart was growing all the time. Her original stem cells morphed into cardiac cells as well as the various organs throughout her body.

"As any child grows, their stem cells continually regenerate, creating new tissue as they go on to adulthood. Of course, adults do have those same stem cells. We maintain our own population as grownups, but they stop working so well. They aren't nearly as plastic and responsive as a baby or a child, whose biology is much more capable of regeneration.

"What we've been able to do is return Emily's stem cells (the ones we used to produce her heart from thin air) to their full *neonate potential*. In essence, we made them suppler, stronger, and, to use her word, more 'powerful.'"

"So, I *am* a superhero!"

"Or at least a super gymnast," said Dr. Aronson with a wink.

Over the next ten years, I used my newfound physical abilities to become a truly remarkable athlete. I never did make my dream of the Olympics, but that's because I stupidly followed my (figurative) heart over my sport when I was in my late teens. After a short-lived and very shortsighted marriage, I bounced around in a series of dead-end jobs.

Don't get me wrong. They paid well. They just didn't inspire me in the same way gymnastics competing once did. I foundered this way into my early 30s, when I learned of a fascinating new medical breakthrough. On accident.

Some friends and I had arranged for a virtual horseback riding retreat. Thousands of miles from Earth and lacking grass and the terrain for a real experience atop a living, breathing animal, we opted for the next best thing: a fully immersive simulation in haptic suits. It enabled my friends and me to experience what it feels like to bound through a meadow at dawn, offering rich olfactory details like the dusty smell of a leather saddle.

Unfortunately, my "horse" was so lifelike—and so rowdy—it threw me, breaking my leg. Talk about embarrassing. How many people do you know who broke their leg on a computer-generated horse?

Right after, I met with a new specialist who told me all about the amazing possibilities of bone regeneration. As the beneficiary of a new heart all those years ago, I was intrigued by the possibilities.

Here's how Dr. Matheson explained it to me. "Since the very first creatures roamed the Earth, the bone life cycle went like this: Creatures have cells that constantly break down their bones. They also have other cells that constantly rejuvenate their bones. That's just natural. Usually, what happens in elderly people who don't exercise, why their bones become brittle, is that this cycle gets interrupted. The restructuring of the bone stops working, especially the longer people or animals live. That is, until now."

Dr. Matheson proceeded to tell me about a new procedure that wouldn't just fix my broken leg but would change the course of my life. You see, I didn't only get the procedure, strengthening my existing bone density, I decided to start my own company around it.

Upon raising venture capital, I started the Revive Project. A cutting-edge biopharma company, we specialize in . . . what else? *Giving people superpowers.* Everyone knows someone (like me) who has a faulty body part. Or two. But why live under such constraints? Not when a solution exists.

Using emerging technologies, we created the first off-world lab for replacing and reviving people's organs, skin, and bones. And we do it based on the same principle Dr. Aronson told me about long ago. The way it works is quite complex, so I will give it to you in the plain English I received at the age of nine. We turn on people's stem cells, tapping into their neonate potential.

This means Grandma doesn't have to worry about a slip and fall that will break her hip (and prematurely end her life). She can replace her parts as they break down. Likewise, people like me who were born with a faulty organ can replace it with a new and updated one, increasing their life expectancy and, more importantly, the quality of that longer life.

In a phrase that Dr. Aronson would understand, we here at the Revive Project turn you into your super selves.

BRIAN ETTINGHAM
Lead Cosmonaut Team. Mars–Jupiter transit, 2088

Theseus's paradox is a thought experiment that raises an interesting question: Can an object that has had all its components replaced remain fundamentally the same object? The famous historian Plutarch raised the philosophical query in *Life of Theseus* from the late first century.

Specifically, Plutarch wished to know if a ship that's been reoutfitted by replacing every single wooden board and plank could still be rightly called the same ship. Here's the ancient text:

> The ship wherein Theseus and the youth of Athens returned from Crete had thirty oars, and was preserved by the Athenians down even to the time of Demetrius Phalereus, for they took away the old planks as they decayed, putting in new and stronger timber in their places, in so much that this ship became a standing example among the philosophers, for the logical question of things that grow; one side holding that the ship remained the same, and the other contending that it was not the same.[1]

I've never been much of a philosopher myself. I'm more a doer than a thinker. Yet the first time I read that somewhere, it reminded me of a novel I once read. Written in the last century, *The Boat of a Million Years* by Poul Anderson had a big influence on my career.

The sci-fi story concerns a dozen men and women who can live forever if no accident befalls them, or no one kills them. Impervious to illness or the ravages of time, their bodies keep replenishing themselves. Their skin never wrinkles. Their hair never thins. Their bones never wilt. At the age of 1,000, they look just as good as they did at 22.

Which makes them the perfect specimens for the exact voyage I am about to take in real life. In the book, these unique humans are perfectly suited to last through the centuries, if not millennia, it requires for interstellar travel.

Unlike in other sci-fi stories, they don't need to be put into cryogenic stasis so their bodies don't break down in transit. Instead, they can use their vast amounts of time to engage in intellectually and physically stimulating activities, from participating in what we might call deep fake conversations with historical figures using artificial intelligence to playing in any number of sports games and activities to keep them in top shape.[2]

Now, at the edge of the 22nd century, my dozen-strong team and I are employing a similar technology. No, I was not born with cells that never age like Anderson's characters. But I benefit from the next best thing: total revivification—courtesy of the Revive Project.

This means I am a lot like Theseus's ship. So is the rest of my team. Yesterday, we said a tearful goodbye to our families. If our mission from Mars to Jupiter is successful, we should reach the giant gas planet in 10 years.

But what if something should happen to us en route? What if someone gets terminally sick? What if someone's heart stops working? What if someone breaks a bone? It's not like we can go see a doctor in the far reaches of space. Not only that, but it's another 10 years back to Mars—if we even make it to Jupiter. That's a long time. Time enough to produce a lot of wear and tear on the human body.

If we ever hope to be real cosmonauts, if we ever hope to leave our galaxy and explore the vast richness of the universe, we must become more like the men and women of *The Boat of a Million Years*. We must replace every one of our cells, even our brains.

When I told that to my sister, she said, "But you won't be you!"

I told her there's a difference between your mind and your brain. The latter is physical and subject to decay and breakdown. The former is fixed and perpetual, but only if one's brain tissue continues to function.

That's when she reminded me of Theseus's paradox. When she was done, she asked, "So, will you still be you?"

I said, "I don't completely know. What *do* I know? There's something out there calling me. Something mysterious. Infinite. And so long as it's out there, I must go to it."

4

DANCING IN THE STARS

Imagine you are the producer of America's hottest new TV program, Dancing in the Stars. *A partnership between Hollywood and the burgeoning space industry, your show is history-making, the first series shot off-world.*

Today you wake up ready for your shoot aboard a private space station. As a seasoned pro, you've coordinated complex productions in dozens of countries, sometimes under nail-biting deadlines. You've also mastered the human challenge, deftly getting incredible performances from entertainers with skyscraper-sized egos.

Undoubtedly, you're at the top of the game.

The one thing you've never come up against before? Microgravity.

What follows is a day-in-the-life dramatization, depicting the challenges and opportunities tomorrow's creatives may soon experience.

6:00 (UTC)

You wake up in a zipped bag strapped to your bulkhead. This is to prevent you from floating off in your sleep. Though you did see Jenna, your director of photography (DP for short), nod off for a second in your last meeting. She looked so *calm* aimlessly drifting.

Not you. You enjoy stability, being in one place when slumbering, although you're pretty sure you slept upside down last night. It's hard to say for sure.

There are no *ups* or *downs* here.

Before doing anything else, you check in with yourself. Before leaving Earth, you studied up on microgravity to prepare. Still, there's nothing like the real thing. Before this, you had always been intrigued by space, but you wouldn't say you exactly geeked out on it. To your surprise, you learned there's a world of difference between microgravity and zero gravity.

> **MA Michael Ashley**
>
> You may be used to Google coordinating calendar invites to account for various time zones. Hovering over Earth in constant motion, you'll see a sunrise and sunset every 90 minutes. It therefore makes little sense to rely on *solar time*. Instead, Coordinated Universal Time will be your timekeeping standard.

Some people think there's no gravity once you leave our atmosphere, but that's wrong. A small but nonetheless real amount of gravity exists everywhere you go in space. As Sir Isaac Newton told us in so many words, gravity is what keeps the planets in orbit around the sun. It even holds the sun in place within our galaxy.

What you're experiencing now as you exit your cozy sleeping bag, floating in your quarters, is free fall. This occurs when an object falls faster and faster, an acceleration due to gravity (albeit a smaller amount than you'd experience on Earth).

This begs the question: How can something fall *around* Earth? NASA has the answer:

Newton developed an experiment to demonstrate this concept. Imagine placing a cannon at the top of a very tall mountain. Once fired, a cannonball falls to Earth. The greater the speed, the farther it will travel before landing. If fired with the proper speed, the cannonball would achieve a state of continuous free-fall around Earth, which we call orbit. The same principle applies to the space station. While objects inside them appear to be floating and motionless, they are actually traveling at the same orbital speed as their spacecraft—17,500 miles per hour (28,000 km per hour)! As you drift along, getting your bearings, you feel a sense of weightlessness.[1]

Unusual but not unenjoyable, weightlessness reminds you of the time your friend got you a gift certificate to a sensory deprivation tank in Venice Beach. Buoyed by warm water chock-full of Epsom salt and sealed in a soundless chamber, you also felt disoriented at first.

Yet as time went by, you found you could float wherever you liked just by using your little finger to gently push off the wall. For a quick second you feel foolish recalling how some crew members laughed at you for trying to *swim* your way around when you first got here.

"That's not how you do it," one said. "And kicking your feet out like that is plain dangerous. You're liable to connect with someone's face."

"Yo. Bloody noses in space are no joke," someone else put it. "Straight up disgusting."

Ever since, you've practiced the subtle art of traversing the station with the slightest touch of the wall. So long as you can connect with a surface using only your finger, the results are dramatic. Applying a modicum of pressure can send you shooting all the way across the room. There's nothing to stop you.

That was fun the first few times you tried it. Now, you're more interested in becoming graceful like your crewmates, controlling your movements and speed. You rely on these newfound abilities to officially start your day.

6:15 (UTC)

Never one to complain about missing a hot shower—long ago you learned the health pluses of cold immersion—getting ready in space is still a culture shock. After all, gravity is what enables us to enjoy the spray from the nozzle whenever we turn on the faucet. In microgravity, water behaves quite differently. Water falls and flows downhill on Earth. In space, it floats in spheres, preventing you from showering or bathing in the ways you're accustomed to.

> **MA** Michael Ashley
>
> On Earth, water stably exists as a liquid, coating much of our beautiful planet—when not frozen or heated to a boil. By contrast, in microgravity, water forms into spherical bubbles or droplets. These float in space not unlike every other object, including you.

Since there's zero chance you can enjoy a hot soak or a shower, you must do the next best thing to freshen this morning: a good, old-fashioned sponge bath. Stripping down in your private curtained area the size of a telephone booth, you clean yourself.

Using specialized liquid soap, you mix it with a small amount of water. Rubbing it on your skin, it forms a lather you remove with a towel. You opt for dry shampoo on your hair. It's better than nothing.

Done with this part of your morning routine, you get out your toothbrush and toothpaste. To your continued amazement, the latter hovers in midair after you let go. Brushing your teeth, your mind drifts back to the subject of water. It's not so pretty to reflect upon where yours came from today. Your space station functions in what's called a *closed-loop system*. All the water needed for each activity—from bathing to drinking—must be recycled. (It's not like you can just pop outside for a cold Perrier!) That means everyone's wastewater must be captured, then reused. This comes from urine, sweat, even your breath's moisture.

Momentarily grossed out, you reconcile your disgust with what you were told by Mission Control before you left: "Any impurities and contaminants get filtered. Not only is the final product potable water suitable for rehydrating food or bathing, but it's also actually *more* purified and cleaner than what most people drink on Earth."

As you ponder this, you get a ping from the studio. Your teleconference is scheduled to begin shortly.

7:00 (UTC)

In October 2021, a Russian film crew returned to Earth after shooting the first movie filmed in space. Actor Yulia Peresild and director Klim Shipenko blasted out of the Russia-leased Baikonur Cosmodrome in Kazakhstan. (Sidenote: cool name for a launchpad!) Joined by Russian cosmonaut Anton Shkaplerov, they ventured to the ISS in a Russian Soyuz to create *The Challenge,* a tale centering around a surgeon sent to the ISS to save an astronaut.

The budget is unknown at present, but it's fair to say it exceeds the typical indie pic. After capturing more than 30 hours of

> **MA Michael Ashley**
>
> The International Space Station is a joint American and Russia collaboration. According to ABC News, "The ISS is divided into two sections: the Russian Orbital Segment operated by Russia and the United States Orbital Segment run by the US. American and Russian astronauts were the first to step inside the ISS in 1998."[2] Costing more than $100 billion, it's the priciest project ever undertaken in space exploration. Comparable in size to a football field, it weighs approximately 450,000 kg and moves at a rate of 28,000 kmph just 420 kilometers above Earth.

footage in 12 days, the trio returned to Earth, beating Elon Musk to the cinematic punch.

No stranger to inflated budgets or dicey shooting conditions, you also see the value of traveling and working lean. That's why you left your typical production team back on Terra—your art director, your costume designer, your script supervisor, all the gaffers and key grips. Instead, you and Jenna, your DP, will wear many hats on this, your first space shoot. She handles lighting, sound, and editing. You're the director responsible for getting great performances from your two stars. (No pun intended.)

The four of you now sit beside a small conference table interfacing with the studio and PR team scattered around Los Angeles and New York. Okay, none of you are really *sitting*. Your two dancers, Mel and Cady, float beside you, sneaking glances now and again at a jaw-dropping view of Earth. Jenna is, as usual, checking and rechecking her equipment. Only upside down.

Of course, it would be much easier to just call the folks back home via phone. NASA has set up an extensive network of antennas over all seven continents to receive just such spacecraft transmissions. These antennas range from the small high-frequency variety for backup communications to the space station to massive 230-foot leviathans capable of communicating with the likes of the *Voyager* spacecraft—now over 10 billion miles away.

However, the PR team isn't interested in a mere voice call.

They see a behind-the-scenes moment like this as pure gold from a public relations standpoint. Most everyone is familiar with singing and dancing reality programs like *The Voice, American Idol, The Masked Singer*, and, of course, *Dancing with the Stars*. They're icons to the public, feel-good fare that still draws vast numbers of eyeballs, even as Must-See TV has largely gone away.

But no one has ever experienced an original TV show shot, pro-duced, and edited in space. *"Dancing in the Stars* is the perfect vehicle to show the people of Earth what's possible," went your orig-inal pitch to the network brass. "Think of all the people who will tune in to watch. There's never been anything like it."

Luckily, you have an excellent track record of producing shows and putting up high Nielsen ratings, otherwise the money folks would never have greenlit your idea. It didn't hurt that you could bring them Mel and Cady, a hot real-life influencer couple with sexual chemistry oozing off the screen. Plucked from their strato-spheric TikTok fame, they bring with them a built-in audience numbering in the tens of millions. Plus, they actually know how to dance.

"We're all super-excited down here," says Steve, head of studio PR, just as soon as your teleconference starts. All of this is being recorded so it can be edited into the final show as part of its unique IP. Sur-rounding Steve on all sides are little video squares populated by all the other media stakeholders invested in getting this first space production right. "Tell us about today's first taping."

"Well, it's not gonna look like all those late '90s and '00s music videos. I'm referring to Britney Spears's 'Oops I Did it Again,'" you explain. "Or that *Men in Black* video."

"Why? I liked those," says Steve.

"I liked them too," you say. "But they

MA **Michael Ashley**

Tomorrow's teleconferencing will rely on relay satellites to send data to the ground. Already, the ISS communi-cates via Tracking and Data Relay Satellites (TDRS), transmitting data to ground stations in New Mexico and Guam.

Relays possess unique advantages concerning com-munications availability. For instance, TDRS at three dif-ferent regions above Earth offer global coverage and near-continuous communi-cations via low-Earth orbit missions and around Earth. Instead of having to wait to pass over a ground station, TDRS users can relay data 24/7, an imperative for busi-nesses requiring constant and dependable connectivity.

don't show you the *reality* of space. That's what makes our show so cool. So very unique."

"Um. What do you mean 'the reality of space'?"

"We're talking about this," says Mel, your male star. He then proceeds to execute insane backflips and contortions defying the laws of physics on Earth.

You turn to Jenna and share a knowing smile. Even *this* moment is scripted. Of course, Steve and the rest of the PR team know all about your microgravity conditions. Hell, they're integral to your show's allure and very concept. But this is Hollywood. Even in space. You're only engaging in these verbal (and physical) theatrics to give context for your show. Weeks from now, when the trailer's finally cut, you can imagine this scene in the glossy commercial that will play for TV audiences, beginning with an expensive Super Bowl promo.

As Jenna breathlessly runs through what else viewers can expect to experience in the pilot episode, you're suddenly struck with your next breakthrough TV concept. You try to stay focused on what Jenna is busy telling the PR folks back home, but your active mind can't help contemplating the next big thing. It's just the way you work as a creative! You make a mental note to pitch Jenna your concept after this concludes so you don't forget it.

9:00 (UTC)

Daily physical training and conditioning starts in one hour with you and the talent. It's grueling, but you must do it.

In the meantime, you've snatched this moment to tell Jenna about your next big reality concept. "Okay. You know how people love to watch shows like Chip and Joanna's *Fixer Upper* and *Nate and Jeremiah by Design*?" you start.

"Uh-huh," says Jenna between sips of water from a tube in her water pack. Watching her, you're once more reminded of your closed-loop space station and how all the liquids on board get recycled.

Pushing that revolting thought from your mind, you continue your pitch. "Why do people like those shows so much?"

"They like looking at nice homes?" Jenna shrugs.

"Well, yes. That's part of it. But there's more. Audiences dig those shows so much because they get a kind of *backstage pass* to creativity."

"I'm not sure I follow."

You realize you must not be explaining this right if Jenna of all people doesn't get it. Time to reframe. "Think about it this way. We're doing *Dancing in the Stars* to show people the future of music and performing, right? It's not just that Mel and Cady are hot—"

"Super-hot," Jenna corrects you.

"Right. *Super-hot.* By setting our series in space, we tap into another viewing demographic, one that might never watch a reality series, especially one on dancing. I'm talking about the sci-fi crowd and the early adopters. The people who share videos of SpaceX launches with their friends. They're gonna want to watch this too, just to see what being in space is like. To get that backstage access."

"Got it. Okay, I'm with you."

"So here's my new idea," you say. "A few years back I went with some friends to this place called Picasso's Palette or something. You get to drink wine with friends while painting. It was really fun and unique—something I hadn't tried before. I wasn't very good, but my friend Kevin was. After a while, I gave up trying to bring my sunset to life with my paintbrush. Instead, I just watched him go at it. Now that space exploration is finally becoming a reality, there are going to be other, *newer* art forms that emerge from microgravity. Or rather, microgravity is going to transform the way we make all kinds of art—not just movies and dancing. Painting too."

"We can make a reality show about future space painters!" Jenna says.

MA Michael Ashley

Astronaut Nicole Stott famously painted watercolors in space that are now displayed at the Smithsonian Air and Space Museum. A veteran of two spaceflights, she served 104 days as a crew member on the ISS and the space shuttle. In an interview on producing art in microgravity, she said the following: "I thought it would be kind of cool to paint in space. What I would do is squeeze just the tiniest little sphere of water out of a drink bag and then shove the brush into that so that it would wick into the brush. I'd then quickly cap the bag off. You just had to be careful you were not squirting water all over the place, or swinging your brush around. The water kind of wicked to the paper, too. It was a nice flow and worked out really easily."[3]

"Right. We'll just need a way to pack it with juicy conflict like *Keeping Up with the Kardashians* so people can't turn away."

Jenna lets go of her water pack. As usual, it stays midair, right where she left it. "I can see it. Totally. Especially the painting part. Long ago, Renaissance painters were the most celebrated artists on Earth. Until inventions like radio and TV—"

"And photographs. That pretty much killed it for them."

"Yes. But what I'm saying is we could do a *Painting 2.0*-type show where we would follow painters doing their thing in microgravity."

"Right!"

Jenna thinks for a moment. "I like the idea of a 'backstage pass' when it comes to exploring stuff in space. But you can go further with microgravity."

"How so?"

"The whole weightlessness thing blows people's minds. Even me. We're so used to heavy things on Earth. Think about your car for a second. It weighs a lot. You're not likely to pick it up and throw it. Unless you're Superman."

"That's true. That's true."

"But all that goes out the window in space," says Jenna. "That means you could also have a backstage pass–type show about 3D printing *giant* things you'd have trouble creating—or moving—on Earth."

You can see she's on to something. "That's right! You could 3D-print monster truck–sized vehicles in microgravity—then film a reality show about new kinds of athletes who throw them back and forth on the moon."

Jenna chuckles. "I'm not sure about that, but you may have something here. It needs work, though."

That doesn't discourage you. What does? Having to go to physical therapy next.

Sunita Williams: Cosmonaut and Space Marathon Runner

On December 9, 2006, Sunita Williams, the second woman of Indian descent to go to space, boarded the shuttle *Discovery* bound for the ISS. The 14th expedition to this space station, it featured a slight twist. Williams did something no other astronaut had ever tried—nor has done since.

She participated in the Boston Marathon while in orbit.

More than 200 miles above Earth, Williams completed the 26.2-mile (42-kilometer) race using the ISS treadmill in 4 hours, 23 minutes, 10 seconds. During this time, she twice circled the globe. "Williams ran under better weather conditions than her Boston counterparts. In Boston, it was 48 degrees with some rain, mist, and wind gusts of 28 mph while station weather was 78 degrees with no wind or rain with 50% humidity," according to NASA.

As it turns out, the Boston Athletic Association issued Williams bib number 140,000, which it sent electronically. As to why competing off-world mattered so much to Williams, she said the following: "I would like to encourage kids to start making physical fitness part of their daily lives. I thought a big goal like a marathon would help get this message out there. In microgravity, both of these things start to go away because we don't use our legs to walk around and don't need the bones and muscles to hold us up under the force of gravity."[4]

12:00 (UTC)

"Use it or lose it," goes the adage about staying healthy. It takes on new meaning in microgravity. According to Don Hagan, director of exercise physiology at Johnson Space Center, "No other activity except eating and sleeping is given that much priority. Two and a half hours each day are devoted to fitness."[5]

That's a big commitment, but there's an important rationale. Terrestrially, our bodies are always pushing back against gravitational forces. Except when sleeping. The bones and muscles of all species, not just our own, evolved over millions of years in support of such resistance.

In microgravity these external stressors all but disappear. Devoid of gravitational pushback, things become too easy for us. Our bodies immediately start losing muscle mass. It's estimated that astronauts who spend four to six months in space, in places like the ISS, can lose bone mass at an alarming rate of 1 to 2 percent monthly.

"Now you don't have to worry about *that* much depletion," you remind Mel and Cady, who are suited up in gym gear. "We'll be heading home next week. But I wouldn't be much of a producer or a director if I didn't insist on daily workouts."

"But I *hate* working out," Cady complains. "We did it yesterday for three hours, and I'm still sore. I almost never hit the gym on Earth."

"Me too," you say. "But this is different. You don't want to come back home all weak and depleted. That can't be good for your performing career."

"I'll be okay," says Cady. "I was already in good shape before I got here. Least that's what they told me in my health eval."

You shake your head and turn to Neil, the ship's physical therapist, for his medical viewpoint. "Doesn't matter, Cady. Being in space is much, much different from Earth. If you don't stand on your bones for a few hours daily—which you're definitely not doing up here—they will suffer."

"Suffer how?" asks Cady, still unconvinced.

"Your bones will release calcium inside the marrow. That lost calcium will come out in your urine, and you won't ever get it back.

> **MA Michael Ashley**
>
> There's yet another health reason to frequently exercise in microgravity. Micro-gravity can also produce something called *orthostatic intolerance*. You've felt this before in the form of light-headedness when standing too quickly. Your body prevents this from happening by increasing your heart rate and blood pressure. If it doesn't, you'll pass out. Lacking gravity and blood volume, astronauts are prone to fainting. Yet one more reason to exercise in space!

"Cady, you're in your 20s," you add. "You're healthy now, but try to picture what it would be like to have brittle bones the rest of your life just because you didn't bother to put in a few hours of exercise."

Cady doesn't need to hear any more. "I get it. Let's just get this over with."

"My thoughts exactly," you say, no more eager than her to work out. "If I could get out of this requirement, believe me, I'd do it too."

Neil then leads you, Mel, and Cady over to a series of fitness equipment. Each is bolted and placed on raised platforms to reduce noise. "Remember these?"

Cady groans. "How can I forget?"

Neil ignores her. "All three machines do something different for you. Let's go through 'em again. This is the Cycle Ergometer."

"Looks just like a bicycle back home," says Mel.

"Right. The main thing here is pedaling. We want to get that heart pumping just as soon as we strap you in."

Neil does just that, securing Cady so that she can begin her cardio. Once she's safely in place, he brings Mel to the treadmill. Mel raises his arms, allowing himself to be harnessed in place. "Like I told you and Cady, you gotta exert pressure on those bones. The treadmill facilitates that by simulating conditions for both walking and jogging."

"Got it," says Mel, already pounding his feet along the rubber surface.

"Now, your turn," Neil indicates for you to follow. "Time for RED."

You mentally break down the acronym. It stands for Resistance Exercise Device. It looks just like the strength-training machines found in gyms everywhere. To use it requires pulling and stretching rubber band–type cords connected to pulleys or, in some cases, vacuum pistons.

"The beauty of this is that it's a *total* body workout," says Neil, securing you to the machine so you don't float away. "We'll start with your legs, doing squats and bending exercises. Then move on to heel raises before we hit your arms."

"Can't wait," you say, already feeling the need for a nap.

3:00–5:00 (UTC)

This is exactly what you do for the next two hours—nap, after you and the talent complete your workouts.

What It's Like to Walk in Space

Astronaut Mike Massimino grew up idolizing Neil Armstrong. He's said he knew what he wished to do with his life after seeing him walk on the moon. The veteran of two space missions, Massimino can also boast participating in four spacewalks to repair the Hubble Space Telescope.

The author of the memoir *Spaceman*, he's since described what it's like being in microgravity—outside the (relatively) safe confines of a space vehicle or station—by comparing it to the feeling of being the least prepared starting pitcher moments before the final game of the World Series.[6]

Furthermore, Massimino explained what he experienced when working on the Hubble Telescope, still tethered to his ship but also awash in the vastness of space:

"From Hubble, you can see the whole thing. You can see the curvature of the Earth. You can see this gigantic, bright blue marble set against the blackness of space, and it's the most magnificent and incredible thing I've ever seen in my life."[7]

5:30 (UTC)

At last, it's showtime. Shooting is about to commence. For the last few hours, Jenna has been setting up her shot, coordinating lighting and arranging a small platform to serve as the dancers' stage. Mel and Cady have exchanged their casual wear (shorts and T-shirts) for a tailored suit and a form-fitting sequined gown.

Members of the crew have also gathered around, eager to see what unfolds today. It's too tricky to teleconference this moment back to Earth, so the PR/media team will just have to see for themselves how the shoot turns out.

For their part, Mel and Cady look polished and relaxed. Consummate professionals, they began terrestrially rehearsing months ago. On their socials, they asked fans to weigh in on which number they should dance to. Obvious entries included David Bowie's "Space Oddity" and Elton John's "Rocket Man."

They ultimately settled on Coldplay's "A Sky Full of Stars." At the start of the first piano notes, Mel enters the frame backwards and upside down. He's moonwalking. Righting himself, he holds his hands out for Cady as he sings along to the song's opening lyrics.

Cady shimmers out to him, spinning like a toy top. Round and round she goes. Offscreen, you anchor Mel's foot so he stays stationary. With just the slightest touch of his finger, he sends Cady spinning. Faster and faster she turns, longer than any whirling dervish could possibly sustain on Earth.

Extending his free leg, Mel throws it behind Cady so when he dips her back her long hair kisses the floor and her bare leg shoots up to the ceiling. Before you can reconcile the unusual sight, Mel has spun out again, using his tethered foot as a fulcrum. Grabbing Cady's arm, he rockets her across the room. All the while she twirls impossibly, flinging arms and legs. As soon as she connects with the wall, she bounces back like a pool ball glancing off a side pocket.

Mel just catches the tips of her fingers. With the slightest flick of the wrist, he summons her back and together they sway to the music.

The tempo picks up as Mel takes Cady by the elbow. Weightlessly he coils her over his head so she flies over his left shoulder, then his right. Looking now like a cowboy with a lasso, he swings Cady overhead, all but throwing her to the very edges of the room before delicately reeling her back in.

As the drums pound to the song's rising climax, you release the strap affixing Mel's foot. Syncing to the beat, he corkscrews up and out with Cady, shooting around and around, faster and faster like a centrifuge. Their hands connected, their eyes twinkling, they look not unlike the rotating disks of beautiful spiraling galaxies.

"And cut!" you yell. "That one's a keeper."

5

YOU'LL SOON EAT THE BEST MEAL OF YOUR LIFE IN SPACE

Duck confit with honey reduction, smoked chile-potato cake, grilled broccolini. Linguine alla pescatora composed of assorted fish, jumbo shrimp, scallops, mussels, clams in a rich wine sauce. Camarones con rajas served on a bed of wild rice, Mexican white prawns, poblano, onions with drizzled garlic butter sauce . . .

No doubt you've encountered sumptuous dishes like these on menus of upscale restaurants. Out with friends or for business, you may have even ordered these delicious entrées, along with mouth-watering appetizers like smoked salmon canapes or socca, a kind of rustic flatbread and pancake rolled into one crispy treat. You may also have paired your meal with a creamy Zinfandel or a smokey Malbec, accentuating the food's rich flavors.

And if you were really fortunate, all these incomparable dishes and spirits weren't your night's *only* highpoints. Rather, they com-bined with other things—who you were with, the occasion you were celebrating—to provide you with the special experience of dining out.

Some venues thrive at intentionally cultivating such indelible memories.

At one end of the spectrum was Tokyo Delve's, a lively LA sushi haunt where singing chefs put on an elaborate sake bomb stage show, encouraging patrons to dance on the tables. It was "true insanity," according to restaurant critic Brant Cox for *The Infatuation*. "If you're looking to publicly swan dive off a table into the arms of a woman you don't know or cry openly during a Britney Spears lip sync, Tokyo Delve's is your spot."[1]

Dining in the Dark, on the other hand, offers a very different experience. Arriving at the restaurant, you're led by friendly serv-ers through pitch blackness to your table. No cell phones, lighters, or flashlights of any kind allowed. Momentarily blinded, you get to feel your way through a seven-course prix fixe meal. Popular entrées include spiced hanger steak and ahi tuna tartare with diced Asian pears, shiso, wonton crisps, and wasabi aioli.

Tasty or not, the food served at Dining in the Dark is not the point. Rather, it's the dance with novelty. As food culture writer Ishay Govender-Ypma explains for finedininglovers.com, "To an eater, and for clarity's sake let's use that grotesquely pompous word—gourmand, the senses of taste, smell, touch, sound, and sight combine to form a package that is vital to the identifica-tion, synthesis, and enjoyment of a dish. What happens when you remove one sense?"[2]

It's a valid question, though not one usually asked as we go about our daily lives, regularly eating and drinking. Being suddenly bereft of sight, something we often take for granted, can produce uncanny gustatory sensations. Lacking the ability to see heightens our other senses, like touch and smell. Most of all, it generates an unforgettable evening in the minds of diners, one they won't soon forget. One they will gush to their friends about.

PLOT TWIST

All that's well and good. *So far.*

But no matter how zany the loftiest chef can imagine a gourmet menu or extreme culinary situation, they are still conscribed by physics. On Earth, chicken wings and nachos don't float. Neither does cutlery. (Okay, that is, unless you happen to be chowing down at Hogwarts Academy.)

Back to nonfiction reality. We are still constrained by gravity and other physical considerations here. Knowing how humanity's future lies in the stars, it's time to contemplate just what fine dining might look like in space. Especially if you take your client to eat in LEO.

First things first; let's be clear, we're not talking about eating freeze-dried strawberries or any other dehydrated foods associated with the NASA program. We're also not discussing Tang, refreshingly sweet as the drink is. Instead, we wish to imagine comestible delights worthy of Michelin stars, served with elegance and care, culminating in the apotheosis of food tourism. In space.

In the following sections, we will describe the problems and possibilities aboard the first future commercial space station. These future stations can be likened to a "mixed-use business park" supporting both orbital research and space tourism—which could include fine dining.

Helping us explore both the profound possibilities and the inherent limitations of this medium is Julian Martinez, executive chef and co-owner of Barbareño, a Michelin-star-recognized Santa Barbara–based eatery at the cutting edge of California cuisine.

The 32-year-old Martinez came up in a highly competitive food scene, apprenticing at marquee establishments like Napa's The French Laundry and The Thomas, San Francisco's Wexler's, and Doc's of the Bay in Oakland. Since opening Barbareño, he's helped launch Cubaneo Restaurant, Quokka Kitchen, Barb's Pies, Bank of Italy Cocktail Trust, and Companio Hospitality Management.

Major publications, including the *New York Times* and *USA Today*, have raved about Martinez's penchant for innovation. A Chef's Corner feature for the *Santa Barbara Independent* describes him this way: "Taking cues from the landscape around him, Martinez forages for native plants such as wood sorrel, bay leaf, and acorns. He also supports hometown purveyors like Winfield Farm, which raises heritage-breed Mangalitsa pigs, known for their fatty, marbled meat. He's able to incorporate Chumash traditions into dishes like oak tagliatelle and grilled chicken, but with a modern spin."[3]

Martinez's reputation for creative thinking—how he utilizes the spaces, nature, and places around him to generate original food creations—makes him a natural fit to explore the realm of the possible for fine interstellar dining.

And for all you would-be space-age foodies, we shall include original recipes in the Appendix crafted by our own Space Chef Martinez and meant for preparation just as soon as civilians begin living—and eating—in the final frontier.

For now, let's start by discussing the one person who can make or break a meal at any restaurant—your server.

INTRODUCING . . . THE SPACE DOCENT

The 1999 comedy *Office Space*, written and directed by Mike Judge (*King of the Hill*), introduced audiences to the archetypical annoying waiter. Or at least reacquainted us. In this case, it was Brian (played by Todd Duffey), an insufferable server at Chotchkie's, a stand-in for chain restaurants like Chili's and Applebee's. You may remember Brian better as the "Flair Guy," who just couldn't wear enough loud buttons and pins on his sporty employee vest.

We first meet Brian as he mimics machine-gun fire to cut into a private conversation that protagonist Peter Gibbons (Ron Livingston) is having with his coworkers about how much he despises his job.

It doesn't get better from there.

Insanely upbeat, Brian oozes phoniness as he runs through his script: "So can I get you gentlemen something more to drink? Or maybe something to nibble on? Some Pizza Shooters, Shrimp Poppers, or Extreme Fajitas?"

"Just coffee," says Gibbons, hoping Brian will buzz off.

Brian barely gets the hint. "Okay. Sounds like a case of the *Mondays*."

Anyone who's ever held a server job knows Brian just cost himself any tip he might have expected by being such a goof. The best servers recognize what they do as a craft, not unlike an art form. To this point, France (where else?) established the Best Waiter in the World Cup in 1961, honoring those service professionals who go above and beyond. It's since expanded to other nations, including Canada, Mexico, and Vietnam.

According to contributor Patrice Novotny writing for *Business Insider*, being able to set the perfect table or flambé a pineapple still isn't enough. Contestants must also answer questions on the fly, like citing Cognac's country of origin and its manner of distillation. "'In

France, customers tend to remember food first with service and ambiance second, a situation that is reversed in English-speaking countries and Asia,' said Patrick Henriroux, a two Michelin-starred chef, also from France. 'People come to restaurants for a range of emotions. We try to provide an overall experience.'"[4]

That last word has particular significance for tomorrow's space server. Or, as we like to call this individual: the *space docent*. A title of prestige denoting a guide, docents are often tasked with escorting and educating visitors to museums and historical sites. We envision future space servers acting similarly, as cosmic chaperones.

Of course, back on Earth, if you encounter a server like Brian the Flair Guy the next time you eat out, you can still take your business elsewhere. Aboard the space station, there's nowhere to go. Visitors and staff are intimately intertwined. This means anyone who assumes this role must necessarily view it through the lens of those restaurant professional servers who compete for Best Waiter in the World.

Put simply, instead of popping by to take drink orders, then food requests, before reappearing with the bill, space docents act as liaisons. Experiencing the same microgravity effects as diners, they too will weightlessly float throughout the space station.

Yet, possessing more experience in this unusual environment, they can act as *culinary mentors*, walking patrons through what can be expected. In this way, they also resemble servers from Dining in the Dark, helping patrons explore the unknown.

Here are but a few more ways space docents will serve as dining guides:

- Showing you how to eat (and drink) upside down, diagonally, sideways, and any other in-between position

- Demonstrating the most practical, yet exciting ways to get the most out of this unusual experience

- Teaching you space dining etiquette (example: how to respect your neighbor's personal space even when you keep crashing into her)

This last point is poignant, as it relates to the other party in this scenario: *you.* Now that we've explored what a new server role might entail, let's look at the futuristic dining experience from a diner standpoint.

FAMILY STYLE—ALL DAY

"In China, most meals are served family-style," explains *World Atlas.* "That means diners are not served individual plates of food as is common in North America. Instead, each diner receives a small bowl of plain rice, and a variety of larger dishes are placed on the communal Lazy Susan for everyone to choose from."[5]

This Lazy Susan approach speaks to informality and conviviality. How many of us have experienced the joy of chopsticking our way through a hodgepodge of Chinese delights from Mongolian beef to walnut shrimp to steamed pot stickers with a fun group of friends?

Similarly, tapas meant for shared consumption are a fixture of many Spanish restaurants. "The relaxed, fun atmosphere they induce stands out as a rare pleasure for Americans accustomed to formal dining experiences," explains our own Space Chef Martinez. Along the same lines, Spanishsabores.com describes the distinction: "One popular definition of the concept explains tapas as small plates, and many people may already associate them with

Spain. But they don't necessarily have to be small—and in fact, the idea of tapas goes beyond the food itself. Here in Spain, tapas are more than just food—they're a social activity meant to be enjoyed in great company."[6]

Okay. Let's be real for a moment. Even if visitors *wanted* a formal meal with assigned seats, correctly placed utensils, and proper attire, this would be an impossible feat.

But as the saying goes, "If you can't beat them, join them." Rather than try to bring patrons or staff down to Earth (no pun intended) with outmoded—and frankly impossible—restrictions in the name of politeness, we have devised our own space etiquette.

Intended to be respectful yet practical—even in the fanciest of cosmic settings—it still recognizes the reality of the situation.

4 Space Eating Tips for Tomorrow's Discerning Diner

1. It's more important to help your neighbor than sit still.
Did the person beside you just float to the ceiling like a character out of *Willy Wonka* and can't seem to get back? Forget about being formal or restrained. Drift over yourself. But before you go, grab a tasty appetizer you can share.

2. Don't bother to learn where anything goes.
On Earth, strict conventions exist concerning which fork to use and how far one's glass should be relative to the plate. All that stuffiness goes out the window in microgravity. Accept informality as a given and even a virtue of space dining.

3. Forks and spoons are useless above.
Most likely you've given little thought day-to-day as to how you pick up food with utensils. (Parents of young kids exempted.) It

might not have occurred to you that you rely on gravity to pierce a steak with a fork's tines. Without this grounding force, it's likewise impossible to spoon soup into your mouth. Our advice? Get comfortable using your hands when eating in the stars.

4. Try to keep your napkin on you.

Remember that bib your parents put on you when you were just a baby to avoid spilling spaghetti all over your shirt? No, we are not saying you should go back to that look. Yet, there's benefit to the bib concept when you're weightless. Especially when you want to wipe your face and can't find anything to do it with.

Maybe in the future some enterprising person will create fashionable bib-like napkins to keep you mess-free without making others think you resemble a tyke in a baby chair. For now, space napkins could be built into a (nailed-down) table, reside on a revolving belt, or be fastened to hang on the walls.

These alternatives sure beat a space bib—no matter how cool it looks.

PERILS OF PRESENTATION

It's time to get into more granular space dining considerations. As stated, silverware and other cutlery are out of the question. Forget about plating an entrée with a beautiful garnish. Both the food and the item it's served on will float away the moment you let go.

Our work-around?

Handheld food. According to Martinez, a major job of (terrestrial) high-end chefs is to curate bites for diners in the way they plate courses. "We can dictate that a crunchy morsel gets eaten in the same bite as something rich and creamy, for instance."

In space, such control vanishes. But it's not impossible to regain it. Martinez has the solution: Dishes could be plated on an edible vessel like a crispy lavash cracker for handheld consumption. "Then, using syringes or squeeze bottles, future space chefs could assemble courses directly on the cracker; for instance, applying a bit of a sweet pureed component, followed by a slightly bitter pureed item, or accompanied with something a little acidic."

Even the shape of the proposed cracker has culinary implications. It will dictate the order in which components will be tasted. Also, the purees can serve as a glue to which a minimal number of toppings could be applied, such as a few pieces of poached shrimp or a kind of textured garnish. "Think about it like a sandwich 2.0," Martinez says. "The idea of something like a typical sandwich—in which multiple textures and flavors can be simultaneously tasted—is not easy to achieve in space. Yet this technique could swing it. It would allow diners to experience a multidimensional course (with varied textures, flavors, and temperatures) in a completely manageable way."

Side note: This method would be especially useful for savory courses. Large pieces of meat and vegetables are not particularly applicable for space consumption, so it would be best to top the purees on the cracker with cured meats, flavorful seafood, preserved vegetables, or aged cheeses, packing in as much flavor and aroma with as few bites as possible.

A WORD ON SPACE PREPARATION

By now it may have occurred to you that the types of cooking we're used to on Earth would be downright dangerous—not to mention impossible—on a private space station. No water boiling. No pan frying. Definitely no firepit barbecues. Does that mean we can't heat up anything to make it tasty? Are we brave new space denizens forever relegated to throwing back dehydrated NASA packets?

Not if Martinez has anything to say about it. "Serving hot and crisp foods would be difficult without the use of sautéing and frying, but there are alternatives. Utilizing super convection, which air fryers and Turbo Chef ovens use, is one solution."

Super convection works by transporting heat from an already hot substance to a cooler one via the motion of one of the substances. Already, so-called air fryers employ this tech. According to thespruceeats.com, "Air fryers don't actually fry. Instead, the food goes into a perforated basket and the machine cooks the food by blowing hot air around it. The force of the air produces a convection effect that cooks and browns the exterior of the food in the basket. As long as the temperature of the air reaches more than around 320 F, breaded foods like frozen chicken tenders or unbreaded starchy items like French fries or tater tots, will in fact turn brown."[7]

While Martinez doubts large pieces of meat like chicken or steak could be air-fried in space, a work-around exists here, too. "Traditional batters would be impossible off-world, as the flours and liquids would float away. However, *lollipop-like* preparations of chicken pate, for instance, would be manageable. These could be coated on Earth using modern crisping ingredients, then stored in vacuum-sealed bags before being air-fried on orbit."

Now, let's explore other preparation techniques and the unprecedented victuals they will produce.

Technique 1: Spherification of Liquids

Enjoying a fine scotch or glass of wine is a staple of fine dining. Yet imbibing liquids in space poses unique challenges. Even if you could pour bourbon into a tumbler, it would escape as little brown droplets. Therefore, any fluid must be encapsulated in some sort of edible or nonedible container.

Spherification holds the key to this challenge, using the former consumable approach. "Created in 2003 by Ferran Adria, spherification is a cooking technique in which a liquid is dropped into a solution to create a thin gel covering around the liquid," writes Angie Bates for Delighted Cooking. "The resulting spheres can then be eaten and produce a burst of liquid flavor in the eater's mouth. Spherification uses concepts based in molecular gastronomy, or the process of cooking by using chemical reactions."[8]

Martinez believes this breakthrough could be applied in many ways, beginning with cocktails. "Still, the last thing you want is a hangover in space. Just imagine the headache or the intense nausea." He therefore recommends using a smaller serving size to create a membrane encasing the liquid.

Both the outer and inner layers would be composed of the beverage—anything from a Cosmic Cosmopolitan to a New-Fashioned Old-Fashioned—but encased together, giving the drinker an explosion of flavor more appealing than drinking out of a water bottle–like vessel or wine box.

Side note: The spherification method could also be used for cold soup courses, like gazpacho or chilled Ramen with soy milk and chili oil.

Technique 2: Liquid Nitrogen Technology

Imagine you wanted to serve something with a creamy texture like a hummus dip or baba ghanoush in space. Would it be impossible? Not if you used our next technique. You may be somewhat familiar with it if you've watched modern TV cooking programs where innovative chefs use it to produce smoky—almost spooky—looking delicacies like fast-dissolving popcorn.

Foodalert.com explains the process of preparing foods and liquids this way. Reaching temperatures well below -300 degrees Fahrenheit, liquid nitrogen can "flash freeze any food it touches. As it boils away, it gives off a dense nitrogen fog that can add atmosphere and drama to food preparation."[9]

Martinez has his own variation on this technique. "Imagine you start with something somewhat creamy, which you could turn into a foam. Back on Earth you could pack it into a puree-filled water balloon. (This would be much easier to store in microgravity rather than puree in a jar, for instance, as it would not need to be opened and spread upon assembly.) To serve, the balloon would be dipped in liquid nitrogen to harden the puree's exterior. The balloon would then be removed, leaving a solid ball with a creamy interior."

Based on this description, we can picture our family-style diners at a business dinner pitching liquid nitrogen balls of spinach dip to each other like Cheetos, catching them in their teeth before chasing them with flatbread. They might also enjoy custard-like dessert such as crème brûlée or pot de crème puff in the same droll manner you could only experience in weightless conditions.

Technique 3: Puffing

Inevitably, meals in space will have fewer gradients of texture and service temperature. Whenever possible, it would be good to introduce such sensations. Consider crunchiness. The pleasing sound we associate with chip snacking would be amplified aboard a space station.

Now how might we infuse cheese with texture? By *puffing* it. We would take something possessing a softer feel and transform it into something more dynamic in space. "Doing so changes the cheese into an edible vessel," Martinez adds. "This could be topped with thick purees and flavorful powders."

Of course, producing something like pizza is not possible in space. Yet, employing the puffing approach, we could use a syringe or squeeze bottle to pipe a thickened tomato sauce onto a chip, dust on powdered charred bread, then use tomato puree to *glue* on toppings like slivered salami.

Also, by drying out fine cheeses, we could preserve them at room temperature rather than taking up valuable refrigeration space— always a plus when your dining area is drastically limited.

Technique 4: Edible Envelopes
Many of the above techniques use encapsulation to contain foods. This means the content inside must be liquified for it to work. But what if you want a way to pack *solid*, textured foods into an edible and flavorful package? It's possible through what we call "the edible envelope."

Think about salad. As we know it now, it would be hard to achieve in space. You can picture floating lettuce, tomatoes, and cucumbers going every which way, not at all connected to a drifting current of olive oil and vinegar.

But this is fixable.

Taking a page from the *Alinea* cookbook created by Chef Grant Achatz (of *Chef's Table* fame), amateur cook Allen Hemberger describes how to prepare the entry entitled Granola in a Rosewater Envelope. "The basic idea is that you make a gelatin-like solid gel of the mixture, blend it (which turns it into something resembling pudding), then dehydrate it. Eight hours later, you have something resembling leathery plastic."[10]

While that might not sound appealing on its face, the significance is huge. Returning to our salad conundrum, we could use this technique to produce a healthy (edible) container stuffed with veggies. As Martinez explains, "Although our 'salad' couldn't be bulky, we could pack finely prepared vegetables, mini-croutons, minispheres of dressing, and diminutive herbs into a flavored gelled 'envelope' for space consumption."

Technique 5: Mochi

Mochi is a versatile ingredient, often employed in Japanese cooking. Traditionally used to wrap pastes and ice creams, nowadays chefs trailblaze disruptive ways to use it. "Mochi (*pronounced MOE-chee*) is a Japanese dessert made of sweet glutinous rice flour or *mochigome*," writes Vanessa Greaves for All Recipes. "Mochi dough is often tinted with green tea powder (matcha) or other food colorings and wrapped around a sweet center to form a small, bite-sized confection with a chewy, smooth, elastic texture. In its traditional form, this kind of Mochi is filled with sweet red bean paste, but in a more modernized version, pastel-colored mochi dough is wrapped around mini scoops of ice cream to make some of the prettiest frozen treats in town."[11]

Martinez likes mochi for another reason. It can be shaped around virtually anything. "Imagine its capabilities for producing fully assembled courses capable of withstanding frozen or refrigerated temperatures. And not just desserts, either. Applying mochi to a fully assembled savory course would allow multiple flavors to come through, yet in a controlled way."

Here's one example of what's possible with savory mochi: An enriched bread could sit at the bottom. Above it would be a foie gras or chicken liver mousse, topped with an acidic cherry gastrique, a dash of smoked sea salt, and toasted spiced hazelnuts, all assembled in a rectangular mold and wrapped with a flavored mochi skin. Yum.

Technique 6: Silicone Molds by Way of Marshmallow Tech

S'mores in space?

Not yet. But we can employ the sweet warm gelatin concoction so many of us know and love in other interesting ways. "First, the use of marshmallows gives us a large substrate that can be warmed," says Martinez. "Most warm foods served in space would have to be purees or bread-like preparations. But marshmallows could be baked in silicone molds and served hot. Second, the marshmallow allows for garnishing, as visually appealing ingredients could be poked into the marshmallow and stick there."

Most people equate marshmallows with desserts, but savory outcomes are also possible. In fact, this technique could apply to soup courses. "We could essentially make the base of a split pea soup, then transform it into a marshmallow," says Martinez. "This could be served on a pumpernickel cracker, garnished with carrot tops and pea tendrils, then draped with a piece of lightly warmed smoked charcuterie."

Reminiscent of the beloved campfire s'more, this method could produce next-level soups with flavors out of this world.

A QUICK WORD ON SPACE FOOD SOURCING

Many, if not all, of the food items suggested could be acquired through standard purveyors. Every recipe is designed with the assumption that most of the prep would be done on Earth, with finishing steps accomplished in microgravity. The key for this to be successful is securing space-bound ingredients in a refrigerator.

Everything destined for the stars would need to be stored in a fastened-down mold or tied down in a robust cooling device for transportation.

🛰

WITH ALL THIS TALK of food and drinking, you may be wondering: Could *I* make these items on my own—before I ever step foot on any private space station? The answer is yes, if you were so inclined.

Below, please find the first-ever sample space menu.

Cocktail Service:
Spherified Negroni with Orange-Rosemary Sugar

Course 1: The Chip as the Dip
Aged Cheddar Puff, Corn Pudding, Tomato Powder

Course 2: Split Pea S'mores
Hot Split Pea Marshmallow with Pumpernickel Cracker and Melting Smoked Lardo

Course 3: Salad in an Envelope
Micro Lettuces with Beet Brunoise, Goat Gouda, Grapefruit Vesicles, Tiny Brioche Croutons, Pistachios, White Balsamic Mini Spheres, Powdered Olive Oil, Orange Blossom Envelope

Course 4: Korean BBQ on a Cracker
Smoked Nori Lavash Cracker Topped with Gochujang Mayo, Kimchi Puree, Shiso Pesto, Gelled Ginger, Topped with Slivers of Raw Wagyu Beef

Course 5: Savory Frozen Mochi

Mochi Cake Filled with Foie Gras Mousse, Brioche, Cherry Gastrique, Smoked Salt, Chili Hazelnuts, Shabazi-Spiced Mochi

Course 6: Dulce de Leche Snowball

Liquid Nitrogen–Frozen Cheesecake Ball with Cajeta Center, Smoked Salt

CHECK OUR APPENDIX in the back of the book for full recipes. Now that we have (hopefully) whetted your appetite, let's talk about the future Silicon Valley of space. Maybe floating around it will help work off some of those cosmic calories!

6

THE SILICON VALLEY
OF SPACE

The Space Station of the Future will usher in the coming Orbital Age.

The private space station is intended to unlock space's potential for more people—especially civilians. It's set to be habitable by the end of the of this decade, with Sierra Space's Dream Chaser® spaceplane providing transportation to and from Earth.

What follows is a dramatization of the unprecedented capabilities we can soon expect.

BY ESTABLISHING A BACKBONE AND CENTRALIZED UTILITIES – INSTEAD OF EVERY VEHICLE NEEDING TO BRING UP THEIR OWN EQUIPMENT – WE REDUCE OVERHEAD.
HMM. DOES THAT ALSO GO FOR OTHER VERTICALS? THINGS LIKE SPACE TOURISM AND ENTERTAINMENT TOO?
ABSOLUTELY. BUT THERE'S YET ANOTHER BENEFIT.
LONG AGO, SILICON VALLEY SAW THE VALUE OF "PROPINQUITY." THEY PURPOSELY DESIGNED CAMPUSES AND BUILDINGS SO THEIR EMPLOYEES WOULD BUMP INTO EACH OTHER.

7

THE NEED FOR SPEED

SPECIAL ISSUE

We here at Hypersonic Revolution *have been covering all things swift, fast, and quick for the last 25 years, ever since Mach 25 became the new industry standard for terrestrial travel. To celebrate just how much life has changed in the last two and a half decades, we are turning over our pages to voices of the Speed Age.*

You will learn firsthand how advances in aviation acceleration shaped the last quarter of a century, improving travel, hospitality, commerce—even world peace. Yes, dear intrepid readers, we have this story and more.

For now, let's buckle in. It's going to be a bumpy ride.

DONALD CATHAY
CEO of Paris to Frisco in 60

A born and bred Parisian, I am 100 percent devoted to our City of Light. All my favorite writers were either born or did their most important work here: Honoré de Balzac, Emile Zola, Alexandre Dumas, Gustave Flaubert, Marcel Proust, and so many others. Of all the many literary quotes describing our magical city, perhaps my favorite comes from Ernest Hemingway, himself an expatriate author who made Paris his home away from home and the setting for so many of his novels: "If you are lucky enough to have lived in Paris as a young man, then wherever you go for the rest of your life, it stays with you, for Paris is a moveable feast."[1]

But to truly appreciate Paris's contributions to art, humanity, architecture, love, and all those things that truly matter, we must go to a surprising source, none other than Anne Rice, herself another Yank author of the enormously popular 20th-century thriller *Interview with the Vampire.*

> Paris was a universe whole and entire unto herself, hollowed and fashioned by history; so she seemed in this age of Napoleon III with her towering buildings, her massive cathedrals, her grand boulevards and ancient winding medieval streets—as vast and indestructible as nature itself. All was embraced by her, by her volatile and enchanted populace thronging the galleries, the theaters, the cafes, giving birth over and over to genius and sanctity, philosophy and war, frivolity and the finest art; so it seemed that if all the world outside her were to sink into darkness, what was fine, what was beautiful, what was essential might there still come to its finest flower.[2]

Now, don't let our confab of so many fine and fancy things related to Paris trick you into believing I'm some sort of Bohemian flower child. Nothing could be further from the truth. At my core, I am a capitalist, one who just happens to be deeply influenced by the arts. This city may have produced in me someone eager to show off our heritage, but it's equally contrived for me to become a carnival barker of sorts: someone who cannot help proselytizing Paris's singular virtues to the globe.

That's how I used my perch as CEO of Paris's premier airline to land so many (frankly, culturally starved) Americans in our capital city during the Hypersonic Revolution. In what I half-jokingly call the World's Greatest Foreign Exchange Program, we imported millions upon millions to the City of Love, a.k.a. the City of Haute Couture, a.k.a. the City of Sun.

Prior to my tenure as airline CEO, it was estimated that a stunning 64 percent of the US citizenry had never gone abroad. Not once had they left familiar shores to marvel at the many treasures and wonders just an ocean away. I credit so many Mach 25 engineers for turning this situation around, in effect, producing nothing short of an aviation miracle. If not even a societal one!

Even now, 25 years later, it boggles the mind how much everything has changed due to the ease and speed of international flight.

WENDY GRODIN
Culture Fit Consultant

My parents' generation were the remote workers, the laptop class. In a lot of ways, they missed out on stuff previous generations enjoyed. How many happy marriages came out of workplace trysts, or for that matter, deep friendships? Once, work was where employees commiserated. They had a term for it: *water cooler chatter*. Management might have frowned on staff gossiping or conferring over the latest sitcom plotline, but their criticism was unfounded.

Plenty of important developments came out of casual hobnobbing.

In fact, there is a term for it now: *sustained proximal networking*. It sounds clinical, like something you read about in a psychological journal, but the premise is basic. People form deep emotional bonds with those in their closest vicinity, those who they spend extensive time with. Poring over decades of social dynamics and patterns, those who study workplace behavior have found that those companies that encouraged such face-to-face relationships tended to have lower employee churn rates—and higher productivity levels.

Looking back now, I get why so many early 21st-century organizations missed the memo on sustained proximal networking. Coming out of the COVID-19 pandemic, there was a sustained emphasis on workplace safety, coupled with a desire for employers to be more flexible about staff working in person. Those businesses that didn't accommodate saw their people revolt. If workers didn't outright quit en masse, they found subtler reasons to show their disapproval, in the form of quiet quitting.

What almost no organizational pundit saw coming was just how the super-sped-up commute would benefit corporate cultures. For years, America's vast highway system was congested, making morning and evening commutes a logistical nightmare. Looking down on mile after mile of clogged freeways, with cars standing bumper to bumper, a visiting alien might wonder, "Is this the best humans can do?"

As they say, hindsight is 20/20. Nowadays, we have that answer. The invention—and subsequent widespread adoption—of hypersonic travel blew apart old ways of commuting. Even more surprising, it catalyzed a golden age of working together.

To appreciate the latter, we need to look at what was happening in businesses across the nation right before COVID-19. It's estimated the typical worker spent 27 minutes one way each day commuting for their work. That's critical time away from family, time that could be better spent working on important projects, exercising, sleeping, and/or doing any number of better things than slogging through gridlock traffic. (And if you lived in the coastal cities like Los Angeles or New York City, it's a safe bet your commute was much longer, sometimes hours.)

It's no wonder then that when many of these commuters finally arrived at work, they felt demoralized—before even starting their day. Sometimes called "the stress that doesn't pay," it led to workplace toxicity, generating feelings of impatience, fatigue, even anger toward one's peers. Remote-based work, once lauded as the solution, did not ease the strain either. Instead, poll after poll showed that workers possessed *increased* stress levels at this time.

This was a surprise.

It flew in the face of conventional understanding. Why would workers report *more* negative feelings toward work and their colleagues once commuting, that odious corporate mainstay, was no longer necessary? The answer flabbergasted even the experts. As it turns out, commuting may have been reviled by the rank and file, but it still provided a greater good: sustained proximal networking. Put simply, commuting was still a terrible chore, but the fruits it produced more than made up for the temporary toll it took on workers.

Without commuting, without sustained proximal networking, workers became more brittle, less sociable. Even their work wins rang hollow. Just how meaningful—or enjoyable—is it to celebrate a closed deal over Zoom? Several years into the remote work craze, it wasn't the employers who turned their backs on the model. It was the workers who reported increased feelings of loss and loneliness.

Mach 25 came online around this time.

It's weird to put it this way, but the Hypersonic Revolution was *the* corporate culture game changer. Or maybe it's when *MiniLines* started.

DEVON WINTERS III
MiniLines VP of Experience

The whole notion of *MiniLines* came out of a fluke, really. Preston Stogel, our CEO, had been keen to bring a kind of Uber-like shuttle to international airports, not just domestically. The thinking was this: You get out of a steamy airport in Bali or Macao, and you need a lift that's *already* there. Already raring to go. No need to queue up for a good 35 minutes along with every Tom, Dick, and his brother, all of you waiting there, breathing over each other 'til the bloody network finally connects you with a bloke on some side street to race across town and ferry you.

MiniLines was meant to fill the gap, pairing airline passengers who just touched down with an autonomous shuttle. It would airlift them home up to 50 miles from the airport where they landed. A first of its kind and damned useful in a pinch, *MiniLines* caught on with knackered passengers and Wall Street tastemakers, straight from the jump. And, like its online forerunners Uber or Google,

MiniLines soon overtook our parlance not just as a noun but as a utility verb, as in "Can you miniline me back to my Islington flat once I grab my luggage?"

But that was only the start. *MiniLines* really exploded as a novel transportation method outright after Preston doubled down on the fledgling acceleration technology. According to his autobiography, *Stogel by Stogel*, the aviation CEO had a vision for transportation no less sweeping than Steve Jobs had for computers, or Henry Ford had for mass-produced automobiles: "I thought to myself, why not reoutfit smaller commuter shuttles like buses and the like with the speed of today's hypersonic airliners?"

And, as they say, the rest is history.

ELLISON DUNBAR
MiniLines Pilot for 10 Years

You have to understand, I came out of the private space industry. I came up flying high-speed, point-to-point spaceplanes with the US Transportation Command. The high point of my career was the day Dream Chaser landed at the Munich Air Show. My co-pilots and I exited the vehicle, grabbed glasses of champagne, and officially toasted the start of winged spaceplane flight.

Before then, spaceplanes launched vertically, requiring massive advances in fuel innovation and rocket propulsion. For years, the horizontal takeoff piece haunted the engineers. At one point, the heads of departments across all the major private spaceflight carriers seemed to collectively throw their hands up in the air as if to say, "Is this really even possible?"

Nowadays, we just smile, content knowing that we flattened the hypersonic barrier. We left it in the dust at the astonishing speed of Mach 25. Call me old-fashioned, but we must go backwards to understand our present.

There weren't a whole lot of folks flying commercially in the early 20th century. Though the Wright Brothers pioneered sustained flight in 1903, by the year 1930, it still wasn't so common for everyday people to fly. The Smithsonian National Air and Space Museum records that only 6,000 souls flew commercially via airplane that year. By 1934, that figure had risen to 450,000. And by the time of the Jet Age, starting in the late '50s with the advent of the 707, there was a 30-fold increase in US airline passengers in just 20 years.

By the early 2000s, millions were flying commercial daily. Still, as a total amount of the global population, the actual number remained quite small. Around this time, Dennis A. Muilenburg, former Boeing CEO, suggested that only about 20 percent of the people on this planet had ever traveled by air. Based on earlier demographics, this translates to roughly 1.5 billion people.

Here's another way to look at the numbers. Not so long ago, back when I took the school bus every morning in my suburban community of Indiana, about 30 kids would ride with me on the 20-minute trek. Beginning with kindergarten and going all the way to my junior high school, that was my daily commute and that of my friends.

We just didn't know any different.

Flash forward several decades, and that's roughly how many people would one day be shuttling in a typical morning commute aboard *MiniLines*. Except we weren't going from one part of a sleepy midwestern town to another as snow flurries dotted our windshield.

For my 30-odd passengers at any given time aboard the *MiniLines*, it was nothing so out of the ordinary to be rocketed from, say, New York to Orlando in less than 30 minutes. A given person would put their earbuds in to watch a show, and before it even ended, they'd be halfway across the country.

And for *MiniLines*' passengers that were traveling in-state, the commute was even faster. "It feels like I only just put on my seatbelt and we're here," riders would sometimes tell me.

MELISSA McDONALD
Spaceliner Systems Designer

B ack in my granddad's day, they had Concorde, the famed luxury airliner capable of hurtling through the air at such dizzying speeds. Only the passenger experience was wanting, to put it gently. Noisy—*really noisy*. There was no mistaking the fact that you were careening across vast distances at unheard-of acceleration. But at what cost? As far as I know, attendants used to even discourage bathroom usage because the plane was just so small.

For those who don't know, hypersonic represents travel that's 5 times the speed of sound. Okay, so how fast is the speed of sound?

It's 762 mph. So 5 times 762 equals 3,810 mph.

That's almost too fast to comprehend, but I'll give it a shot. There was once a baseball player named Cool Papa Bell who equally defied the limits of speed—but in human form. He played in what was called the Negro Leagues from 1922 to 1946, right before Jackie Robinson broke the color barrier into Major League Baseball.

Just how fast was Cool Papa Bell? Legend went that "Bell was so quick he could turn out the light and be in bed before the room went dark."

I don't know about all that. It was before my time. But I did grow up hearing tales about the Concorde, a jet boasting a tagline reminiscent of Bell: "Arrive before you leave."

Urbane, sophisticated, Concorde represented the high-water mark of aviation and elegance, delivering well-heeled civilian passengers westward across the Atlantic in a mere three hours. Flying at twice the speed of sound and quicker than a rifle bullet, by the time the flight attendant tipped champagne into your crystal glass, you'd already covered 26 miles.

Too bad they'd already retired Concorde in 2003, the year before I was born. I never got to "Fly Me to the Moon" as Frank "Ol' Blue Eyes" Sinatra once sang, himself a high-flying Concorde aficionado. But that didn't mean I couldn't take a page from its playbook.

When commercial airlines all switched to hypersonic travel, it was up to engineers like me to design the initial fleets. Recalling the Concorde's aviation prowess, I submitted plans for aircraft with adjustable droop noses, revamped brake systems, delta-shaped wings, and—my own personal touch—a roomy fuselage permitting attendants and passengers alike space to breathe.

Of course, we all knew the passenger experience would transform with the emergence of *boomless* supersonic craft. Travel costs fell while an ancillary convenience industry took flight. Picking up where Concorde left off, companies like Stow-It-All formed to provide passengers with those amenities they'd lost—or never enjoyed in the first place: expanded legroom, in-flight sleeping options, gourmet selections cooked to order, including fresh-caught seafood. Eventually, virtual reality even gained a foothold as enterprising airlines like my own employer began converting unused portions of the plane into pods for physical exercise while still in flight.

Anyone who lived through the first quarter of the 21st century saw a once viable global supply chain buckle under the incredible strain. Thankfully, space freight advances, coupled with hypersonic transportation, reversed more than a decade's worth of logistical stumbles and missteps.

More on the problem. For years, a withering network of ground transport struggled, epitomized by rusting trains built more than a century before. Packed with critical shipments, these ancient trains hobbled along, working alongside a shoddy patchwork of ships, trucks, and drones.

In the 2020s, frequent supply chain breakdowns occurred even as newer RFID and GPS technologies rolled out. Despite such advances, it was sometimes impossible to track where items went or if they were ever shipped at all. Meanwhile, unrepaired train tracks began showing deep signs of wear due to nearly continual usage over so many decades.

Things came to a head as the US manufacturing base finally admitted it could not compete on the world stage using a transit hub designed long before AI was the primary logistics driver. Rather than try to upgrade miles and miles of track, as well as the trains and vehicles transporting goods long-distance, the federal government passed the Space Transport Act.

Viewed as a one-of-a-kind moonshot project with potential to transform daily life, it called on the nation to phase out reliance on railroad train delivery, favoring *space freight technologies*, delivering goods at hypersonic speeds.

These days, it's hard to overstate just how dramatic the change was to both domestic and foreign shipping. People used to joke,

"Your check's in the mail," alluding to how long it once took to send anything, even a letter.

We don't have those problems anymore. We're all beneficiaries of a hypersonic logistic system capable of blasting a package out anywhere on Earth—or in LEO—in less than three hours.

Goodbye, shipping delays. Forever.

DENIS EZTHER
CEO/Founder of Builder 10,000

My dad grew up in the "Don't bug me, I'm working" construction era. Back then it would sometimes take a year—or more—to finish a frame house. Crazy, right? It's even more nuts to think about how the first colonists here in our New World could erect an entire log cabin in days.

Delays in the modern era came from poor distribution and poor access. The desired timber might be available for a new job, but due to costs, it wouldn't be feasible to transport materials to a job site for months. Also, there was always the problem of moving workers, materials, and tools across long distances to where they were required.

Whenever I learned of such delays on a job I was working, it'd make my blood boil. All that waiting meant my people wouldn't get paid on time, that all those construction projects would just sit there.

Not anymore.

My company Builder 10,000 just completed our contract to develop a 600-unit subdivision on the moon. Getting materials and tools to the site wasn't half the challenge I had expected. Space freight teams managed all the transport for us at hypersonic speed.

As a matter of fact, just as soon as I clicked my approval on items, delivery probes would blast off, in some cases beating my own crew members to the job.

A student of history, I liken our era to the days after the US built the first continental railroad. It's hard for modern folks to imagine there was once a time before people could just order items like batteries or tents, that our ancestors had to get everything they wanted from nature or by trading with their neighbor. It's going to be equally as hard or harder for future generations to imagine a time before hypersonic flight could deliver products and goods at extreme velocity.

Still, judging by the early successes of lunar building, it may not be as hard as some have predicted to one day build on Mars. For one thing, we know it's possible to move materials at a breakneck pace. Now that there are large-scale factories in low-Earth orbit, it's not at all utter fantasy to think we could begin producing a whole new Martian civilization.

COL. ANTHONY ESTEBAN
Upper Earth Security

The day terrorists launched a ballistic missile at the United States was one for the ages, that's for sure. My great-grandparents grew up thinking we were but a few cocktails away from nuclear Armageddon. Give Khrushchev a White Russian or whatever the man drank—I don't know what—and we could be facing down another Cuban Missile Crisis. Or worse.

The nukes I'm talking about came from the debacle in Afghanistan. Or at least that's what I've been told. They could just as easily have been bought off some Soviet turncoat for the price of a steak and lobster. Don't ask me. That's not the important part of this story.

You see, while people make plans, God laughs. Or something like that. As it happens, an international contingent of space-pioneering nations, including the US, the UK, Canada, Australia, Japan, New Zealand, France, and Germany, had recently gathered for a summit to study the feasibility of permanent Martian settlements.

This was a full two years after the same bloc came together to develop an international lunar zone. What a name!

It smacks of bureaucratic red tape, the kind of place you might've huddled in Iraq to avoid insurgents lobbing IEDs at your convoy. In reality? It was closer to Levittown, USA: a cul-de-sac of pill-shaped houses to accommodate those first lunar settlers.

Supported by continual hypersonic transit of supplies, including modular building materials and hydroponic gardening setups, it had enough self-generating tanks to keep all those gutsy first gens flush with oxygen for the arduous months, if not years, ahead.

That's where my mind was when the whole goddamn s---storm broke loose. The countdown to nuclear impact ticked off in my AR peripherals. Meanwhile, the State Department was screaming bloody murder in my face—all wide-eyed about the imminent threat. As for me, I was coolly detached, watching the numbers count down to boom.

Except that wasn't really how I intended to spend my so-called last moments on Earth. The augmented reality feature display in my right eye corresponded with the time until impact, yes. But I could also see out of my left eye the time until a different type of collision.

Whoever these terrorists are—or, more accurately, were—they weren't clued into recent propulsion advances. Their missile o' doom might as well have been a kindergartner standing up as he pedals with all 50 pounds of his might to get to school on time—compared to our knockout missile.

The latter in this situation was more akin to a Porsche 911 racing the same kindergartner. Just no chance in hell our hypersonic troubleshooter wouldn't overtake the errant nuke. No matter how the latter got the drop on us.

The whole thing was done in less than three minutes.

The subsequent press briefing outlasted the whole fiasco by a good hour. You see, when a supersonic missile goes up against a hypersonic *destroyer* missile, there's just no contest. That's due to the basic fact that hypersonic speeds correspond to very high supersonic speeds. In plain English, we are talking about the speed of sound versus five times the speed of sound.

That nuke never had a prayer.

PEGGY DILLWORTH
US Senator

I f my wonderful constituents had their say, my reflections on this milestone would concern the many achievements we can be so proud of having created together. I'm talking about the contract with Barnes Jewish we secured, enabling first responders to arrive at any medical crisis no matter how remote within just minutes. Oops. There I go again, thinking like the career public servant I am. I really must rein that part of me in. It's almost primary season.

All kidding aside, my contribution to this retrospective is more personal. As the only child of two loving parents who sadly passed before my wedding day, I have come to appreciate—okay, maybe a better word is *obsess*—over family time. All the other young moms in my little circle didn't seem to fuss over details as much as I did. Or, at least, they didn't make it as obvious as me. They didn't insist on at least a dozen photos of their own Baby Jim blowing out his first candles. (In actuality it was closer to two dozen, but who's counting?)

They also didn't beat themselves up for not spending every waking moment with their littles. Maybe they just blocked it out of their heads that their little Jim and their little Issy would only be this small and this adorable and have that little lisp for so long. Well, perhaps that's exactly what all those young mothers like me thought. I was just too preoccupied with my own life and my own kids to realize it.

Okay, now that we have completely and utterly established my (neurotic) attachment to my kids, it's the part of my tale where I connect it to the subject at hand. As I already alluded to, a secret monitor in my mom-brain was forever crunching numbers. Numbers concerning time. Or rather, *lost* time. A kind of internal ticker would alert me that it was nearing 8 p.m., the kiddos' bedtime, and I hadn't spent more than a few minutes with them outside of a hasty breakfast.

Signal received, I would stop whatever I was doing to make sure I squeezed in at least one precious hour with my babies before they went down. Often, the person who suffered the most in this situation would be my PA, Lulu, who'd have to make any excuse she could for me at whatever function we were at, while I snuck out.

That was back when I was in local politics.

The year Jim turned six and Issy four, I was elected US senator. Temporarily residing in Washington, DC, and being hundreds of miles from our home in Denver didn't mitigate my internal alarm in the slightest. If anything, it made it more sensitive to the absence of my babies.

And let's face it. They were babies back then. My oldest had just started first grade. His little sister still wore pull-ups to bed. And here I was expected to live and work halfway across the country. My husband and I were able to weather the time apart. We met in politics, after all. We both interned for the same senator in what feels like a lifetime ago. So, my Martin understands sacrifices between us. And he's okay with them. Or, at least, he does a good job of pretending he's all right with everything.

He does put on a brave face about it all. I saw the real one the first time I took myshuttle. Poor Martin. I never even told him my new position came with such transportation perks. He didn't know I could sail away from work on a hypersonic *personal jet*. He had no idea something existed that could rocket me back home in less than an hour.

So, that first time I showed up unannounced at bedtime, shock filled Martin's eyes. "How'd you get here so fast? We just spoke on the phone . . ."

Dressed in those cozy pajamas of his that I love, he had an arm over both children's shoulders as he read *Goodnight Moon* to them. After we put the kids down together, I took him outside to see the self-driving aerial vehicle just big enough to fit one passenger.

"You wouldn't believe how fast she flies, darling."

"How fast?" he asked.

"You feel like pancakes for breakfast? I can get 'em from New York and be back before your head hits the pillow."

While this world—and the ones we are seeding on the moon and Mars—can point to all kinds of financial reasons why hypersonic flight is so important or why it's good for the environment or global security, I have my own priorities. For me, it's one word: time.

Flying so swiftly across the darkened sky that it's not entirely clear if you're in this atmosphere or the next would be little more than a physical rush if not for what it enables. Traveling at such unimaginable velocity can't help but change your perception of what it means to be human. Empowered, emboldened, it's as if you take on an exoskeleton, protecting you against everything that's not you, that could halt you or slow you down.

And that's not a bad feeling.

But the real superpower I feel from so much speed is the force of time. I am clawing back the hours and minutes from eternity. Here, that time will stay with me just a little longer. These precious moments with my family are mine and mine alone, in my tight grasp just a moment or two longer.

That's enough for me.

8

HUMAN SPACE TRAINING TAKES FLIGHT

The following is based on real events, slated for production in Space Hollywood.

EXT. SPACE - TIME UNCERTAIN

We open on a vast inky darkness dotted with jeweled specks of STARS like diamonds. A cigar-shaped SPACESHIP swims into view. It's not much bigger than a 20th-century Panzer tank.

                                        DISSOLVE TO:

INT. SPACESHIP - CONTINUOUS

Push in on the BRIDGE.

Sophisticated controls line smooth metallic walls. It's tech centuries beyond anything NASA has produced.

We linger at a vacant CAPTAIN's CHAIR before going to . . .

INT. SPACESHIP (HALLWAY) - CONTINUOUS

Various scientific OBJECTS of unknown usage float in
microgravity.

Blinking red sensors exude warmth in the otherwise
cool blue light.

INT. SPACESHIP (QUARTERS) - CONTINUOUS

At last, we reach the bedroom.

A floor-to-ceiling PANORAMIC WINDOW dwarfs a TETHERED
SLEEPING BAG and small DRESSER hovering inches from
the ground. The window offers a breathtaking view of
the galaxy.

CAPTAIN ELWIN MATER (young but with timeless
features) blinks his eyes open. He's an incredible
human specimen: tall, muscular, healthy, and handsome.

Yawning, he releases a RESTRAINT BELT running the
length of his chest.

> **ELWIN**
> Lights.

Instantly, a soft glow not unlike the sun brightens
the room. Then we hear the sound of a female voice.

The walls light up with color as an AI speaks to Elwin.

> **PRECEPTOR JYN**
> Finally. As my grandma would say, "You're
> sleeping away your life."

Elwin scoffs as he checks his NOVAWATCH (Think: a
Fitbit on steroids).

> **ELWIN**
> I only got 8 hours last night. And you
> don't have a grandma, Jyn.

                    PRECEPTOR JYN
                      (mock hurt)
        Ouch. Not everyone is blessed with carbon-
        based physicality.

Elwin launches into his morning stretching routine.

                        ELWIN
        Touché. Speaking of which, can you assume
        some form? You know I don't dig the whole
        omnipresent thing.

                    PRECEPTOR JYN
        I thought you'd never ask.

Instantly, a 3D hologram, the stunning likeness of a
frightening KING COBRA, slithers across the floor.

                    PRECEPTOR JYN
                   (now in snake form)
        Ready for your lesson?

                        ELWIN
                       (deadpan)
        Not on your life.

Preceptor Jyn morphs into a MASTODON nearly filling
every inch of the room.

                        ELWIN
                   (still unfazed)
        Nope.

The mastodon vanishes. In its place a small BOWL
appears with Preceptor Jyn inside, this time as a
talking GOLDFISH.

                    PRECEPTOR JYN
            We really should get started with your
            lesson. We're burning daylight.

                        ELWIN
            Then be something—or *someone*—normal.

Preceptor Jyn becomes an Earthling FEMALE about his
age.

                    PRECEPTOR JYN
                        (pouting)
            Humans are *so* boring.

                        ELWIN
            Watch it. We created you.

Now in human form, Preceptor Jyn takes the
opportunity to help herself to the ESPRESSO MACHINE
in the corner.

                    PRECEPTOR JYN
            Want a coffee? We got a lot to cover today.

Elwin takes her up on the offer.

                        ELWIN
            I still don't get why you bother. Does
            caffeine even affect you?

Preceptor Jyn coolly sips her cup with affectation, one
elegant pinky in the air.

                    PRECEPTOR JYN
            My circuity affords me the luxury of taste.
            You should be so lucky.

Rolling his eyes, Elwin heads for the bridge until
Preceptor Jyn blocks his path.

> **PRECEPTOR JYN**
> There's nothing to monitor. The *Nav's* got
> this.

For the first time, apprehension flickers across Elwin's
face. *Why's he worried?*

> **ELWIN**
> So, it's really happening.

> **PRECEPTOR JYN**
> In three hours, you'll be . . . home.

Elwin's face hardens.

> **ELWIN**
> Not my home. I haven't been there in ages.

> **PRECEPTOR JYN**
> Yes, but your parents were born on Earth.
> You were born there too. That makes it
> your home.

Elwin tries to look unconcerned.

> **ELWIN**
> Fine. Let's just get into it. Where'd we
> stop yesterday?

The moment he says this, Elwin's quarters transform
into what looks like a FRONTIER CLASSROOM from the
1800s.

Preceptor Jyn dons traditional SCHOOLTEACHER GARB to fit
the part. Elwin sits at an old-fashioned WOODEN DESK.

> **PRECEPTOR JYN**
> You were struggling to understand The
> Decline.

ELWIN

I wasn't *struggling*. It's just hard to wrap
my head around it.

PRECEPTOR JYN

It was a bit before your time.

Preceptor Jyn drops the smart-aleck act. Patient
instructor that she is, she presents a 3D vision like
a TV monitor.

CLOSE ON:

HOLOSCREEN

We glimpse images of our current era and the not-so-
distant future.

Images of overweight, unhealthy HUMANS stare back at
Elwin. The contrast between them and him couldn't be
starker.

Elwin is an amazing human specimen——whereas these
people look sickly, unhappy, and unwell. (Imagine the
difference between fit people and the stout passengers
in *WALL-E*.)

PRECEPTOR JYN

You're looking at real pictures of humans
circa 2030, some decades before you were
born.

ELWIN

They're so . . . *soft*-looking. Weak.

PRECEPTOR JYN

Their minds weren't much better. After
TV's invention, then the internet, literacy
rates collapsed.

                    ELWIN
        What were they before?

Holoscreen shows an image of early American SETTLERS.

                PRECEPTOR JYN
        The early colonists were avid readers.
        Long before video games and streaming
        came along, America enjoyed a robust print
        culture. People read newspapers daily.
        They didn't need their news condensed to
        140-character soundbites.

Holoscreen shows PEOPLE dressed in old-fashioned
clothes reading by candlelight in LOG CABINS.

                PRECEPTOR JYN
        Civic life required an informed public
        capable of understanding the New World and
        its demands. Thomas Paine's *Common Sense*,
        published in 1776, sold 400,000 copies.

                    ELWIN
        Is that a lot?

                PRECEPTOR JYN
        The equivalent in the 20th century would
        be about 30 million copies.

Elwin points to COLONISTS on the Holoscreen.

                    ELWIN
        They look hardy too.

Holoscreen shows Columbus's SHIPS sailing across the
Atlantic.

PRECEPTOR JYN

That's because they were descendants of
explorers who traversed the globe seeking
new opportunities, new lives. Back then,
journeys across the ocean could take
months—even years.

ELWIN

Years? Just across the ocean? That's only
5,000 miles. We can now cover *millions* of
miles in seconds.

PRECEPTOR JYN

You wondered how people became "soft" in
The Decline. Sailors—the prototype for
astronauts—only consumed 3,000 calories a
day. If they were *lucky*. They also worked
their muscles constantly, enabling them to
withstand extreme conditions.

Holoscreen image of a London TAXI DRIVER circa 1960s.

PRECEPTOR JYN

There's something else. Before self-driving
cars, cabbies had to memorize labyrinthine
street grids. London taxi drivers called
it "The Knowledge"—akin to a black belt or
college degree. This ballooned the size of
their brains, especially their hippocampus.

ELWIN

Making them smarter?

PRECEPTOR JYN

Smarter, hardier, more resilient.

Elwin paces the room, thinking.

> **ELWIN**
> Yesterday, you told me in The Decline, people's IQ fell. So did men's testosterone levels. Birth rates declined. In every possible way, humans were a shell of their former selves.

> **PRECEPTOR JYN**
> I didn't put it so eloquently, but yes. That's accurate.

> **ELWIN**
> What turned things around?

Holoscreen shows The Human Flight School.

> **ELWIN**
> *Astronaut training?*

Holoscreen shows the "Right Stuff" ASTRONAUTS.

> **PRECEPTOR JYN**
> If space training was just confined to a handful of professional spacefarers, it wouldn't have changed the trajectory of your race.
> > (beat)
> But you're on the right path.

Holoscreen shows an aquarium-sized TANK with swimming DIVERS.

> **PRECEPTOR JYN**
> But first we must talk about pain.

> **ELWIN**
> Ouch!

Preceptor Jyn touches Elwin with her finger, shocking him. He collapses to the floor in agony.

> **ELWIN**
> What'd you do that for?

> **PRECEPTOR JYN**
> They say, "No pain, no gain."

> **ELWIN**
> Who says that?

> **PRECEPTOR JYN**
> Adversity builds character. It strengthens
> muscles, grows minds.

Before he can argue back, Preceptor Jyn leads Elwin
into the 3D aquarium tank simulator.

> CUT TO:

INT. AQUARIUM TRAINING TANK (SIMULATION) - CONTINUOUS
Underwater, Elwin and Preceptor Jyn wear pressurized
SUITS.

Elwin desperately struggles to maneuver in his, but
Preceptor Jyn takes it in stride.

> **PRECEPTOR JYN**
> (speaking via headset)
> Perhaps the most physically grueling
> training was underwater. Not because you're
> underwater either—many people scuba dive.
> It's because you're in a heavy, pressurized
> suit underwater.

Elwin strains to move his hands, fingers, and arms.

> **ELWIN**
> (panting through headphones)
> How . . . is . . . this . . . so . . .
> easy . . . for . . . you?

> PRECEPTOR JYN
>
> I reset your physical capabilities in the
> simulation. You're operating at the level
> of a late 20th-century space worker.

CUT TO:

INT. V.R. SPACESHIP - CONTINUOUS

Free from the water tank, Elwin sits in the JUMP
SEAT, piloting a DREAM CHASER in this new simulation.

Preceptor Jyn sits beside him.

> PRECEPTOR JYN
>
> You're on manual. Try not to screw up the
> orbital mechanics, K?

Using the hand controls, Elwin tries to fly the
spaceplane.

He attempts to traverse a straight line. It doesn't go
so well.

> ELWIN
>
> This is so hard. People used to have to do
> this themselves?

> PRECEPTOR JYN
>
> Just try rendezvousing with that vehicle
> there.

Elwin fails in his attempt at orbital mechanics,
crashing into the other ship.

The explosion takes us to——

EXT. WILDERNESS (SOMEWHERE IN NORTH AMERICA) - DAY

Though Elwin wears heavy GEAR, it's clear from his
ruddy cheeks and foggy breath that he's freezing.

Beside him in a warm parka looking carefree is
Preceptor Jyn.

Elwin tries to start a fire with wet MATCHES.

> **PRECEPTOR JYN**
> We talked about space workers earlier.
> They differed from astronauts. Almost
> none had military or outdoor/survivalist
> training. They weren't space professionals,
> but they weren't tourists, either. They
> were civilians—professional researchers,
> scientists, manufacturers—put in extreme
> situations.

> **ELWIN**
> (between chattering teeth; re: starting the fire)
> You could help here, you know?

> **PRECEPTOR JYN**
> I'd rather watch you. Builds resilience,
> you know?

Elwin finally lights the match. It promptly goes out.

> **ELWIN**
> Damn.

> **PRECEPTOR JYN**
> Camping is a lot like space worker
> training in many ways.

> **ELWIN**
> (shivering)
> Because it's cold like space?

> **PRECEPTOR JYN**
> That's part of it.

CUT TO:

EXT. WILDERNESS (TENT) - NIGHT

Elwin and Preceptor Jyn lie in sleeping beds in a cramped tent simulation.

It's so cold they can see their breath.

> **PRECEPTOR JYN**
> People who really love adventure, who adore going on camping trips like this, were better equipped for roughing it in space.

> **ELWIN**
> Can you excuse me?

> **PRECEPTOR JYN**
> (playing dumb)
> What's the matter?

Elwin fumbles with the tent ZIPPER.

> **ELWIN**
> Um, nature calls.

Elwin stumbles out, tripping over the tent flap. We can faintly hear the sound of him making water outside.

> **PRECEPTOR JYN**
> (ignoring the urinating)
> Hardy, adventurous people are less touchy about going to the bathroom in front of others. They're also okay with not showering. For days. Even weeks.

CUT TO:

INT. HUMAN SPACEFLIGHT CENTER - TIME UNCERTAIN

Elwin and Preceptor Jyn step inside a simulation.

They watch as a crew of newbie TRAINEES in microgravity navigate the bathroom situation for the first time.

A tough, floating, no-nonsense INSTRUCTOR looks on as he floats beside them.

> ### INSTRUCTOR
> Okay, cadets. Today's all about dispensing with your modesty. Up on our station, everyone's gonna get real familiar with each other.
> > (beat)
>
> *Real familiar.*

FEMALE TRAINEE picks up a long HOSE attached to a TOILET.

> ### FEMALE TRAINEE
> Is this what I think it is?

> ### INSTRUCTOR
> Up there in the stars, *everything* floats, right? You float. Your food floats. Your . . . waste floats. You can't even sit down to pee.

> ### FEMALE TRAINEE
> We have to be strapped onto a toilet?

> ### INSTRUCTOR
> No, we don't use straps. A handhold or foothold will work just fine.

Instructor takes the hose from Female Trainee.

> **INSTRUCTOR**
> Just remember: Anything that comes outta
> your body *will* float. People don't really
> think about that. They also don't really
> think about what might happen should their
> waste not be accounted for. Should it float
> away, if you get my meaning . . .

Horrified, Female Trainee makes a gagging sound.

> **INSTRUCTOR**
> It's like forgetting to take out the trash.
> *But worse*. There's no escape hatch in
> space. You and everything else are stuck.
> Together.

Preceptor Jyn turns to Elwin standing outside this
tableau but watching it in real time.

> **PRECEPTOR JYN**
> You're spoiled, Elwin. Back in the old
> days, *everyone* had to be very clean in
> space. Very methodical about their hygiene.
> It's like camping in the woods. You can't
> just leave some food—or, in this case,
> waste—out. In the woods, a bear could get
> into it. In space, whatever doesn't get
> properly put away lingers. Forever.

We push in on Female Trainee as if she just heard
this.

> **FEMALE TRAINEE**
> Gross.

                                        CUT TO:

INT. EARTH CAVES - TIME UNCERTAIN - CONTINUOUS

Another simulation: Elwin and Preceptor Jyn go
spelunking in a very tight cavern space.

There is barely enough room for them to crawl through
the wet and darkness.

> **PRECEPTOR JYN**
> Having to be extra careful about one's
> hygiene might seem trivial to our
> discussion, but it's not.

Elwin struggles to move in the enclosed area. It's
damn near impossible.

> **ELWIN**
> Can we pick this up later? I'm a little
> preoccupied.

Preceptor Jyn ignores him.

> **PRECEPTOR JYN**
> Needing to be continually mindful about
> how one eats, drinks, and does everything
> else deeply affected that first generation
> of space workers, astronauts, and tourists.
> The space pioneers. All those hardships
> transformed the people involved. It shaved
> down pounds, built muscles, developed
> latent physical abilities. In short, it
> turned around The Decline.

Elwin isn't listening. He's trying not to lose his
mind in this claustrophobic nightmare.

> **ELWIN**
> That's nice.

Perfectly at ease, Preceptor Jyn goes on.

> **PRECEPTOR JYN**
>
> Of course, these were only minor creature
> comfort inconveniences. The real
> training—*and personal growth*—occurred
> mentally. Getting comfortable with being
> uncomfortable.

By now, Elwin is nearly clawing the walls to get out.
Preceptor Jyn remains calm as a cucumber.

> **PRECEPTOR JYN**
>
> There's another corollary. Back in the
> seafaring days of Captain James Cook
> aboard the HMS *Endeavour* in the 1700s, life
> was also tough. Hellish, even.

> **ELWIN**
>
> Hellish.

> **PRECEPTOR JYN**
>
> Cook—who, by the way, was the inspiration
> for *Star Trek*'s Captain Kirk—was
> responsible for making big advances in
> scientific knowledge in what we call the
> Age of Reason. But it wasn't easy. It was
> miserable. Cook's ship was only 109 feet
> long and 29 feet wide.

Elwin tries to concentrate on Preceptor Jyn to get
his mind off his claustrophobia.

> **ELWIN**
>
> That's about the length of three school
> buses.

PRECEPTOR JYN

Yes, but remember, Cook's ship housed 94
men, 10 cannons, 12 tons of iron ballast,
plus drinking water, chickens, a goat, two
greyhounds. And some pigs.

ELWIN

I'm beginning to feel as cramped as they
must have felt.

PRECEPTOR JYN

(ignoring him)

At least Cook's men got fresh air. And the
chance to stretch their legs now and then.
Those first-gen space pioneers were confined
to their metallic surroundings. Can you
imagine the claustrophobia? It must have
felt like the walls were caving in on
them.

ELWIN

(sarcastic)

I can't even imagine.

PRECEPTOR JYN

On the other hand, they trained for such
extreme conditions. And used their pain
and discomfort as mechanisms for growth.
No mental and emotional weaklings there.

(beat)

You doing okay, Elwin? You're awfully quiet
back there.

Elwin just stares daggers back at her.

CUT TO:

INT. SUBURBAN RESIDENCE - DAY

Visibly relieved to be free from the cave, Elwin accompanies Preceptor Jyn in a new simulation.

This time they have entered a tableau of a home circa 2025.

They look on as MOM, DAD, and two KIDS (5 and 7) sit at the KITCHEN TABLE conducting what looks to be a family meeting.

(They can't hear Preceptor Jyn or Elwin.)

> **PRECEPTOR JYN**
> First-gen space pioneers stayed up there for varying lengths of time. Professional astronauts: Up to a year. Space workers: A few months. Tourists: A couple weeks.

> **ELWIN**
> There must've been a lot of emotional toll involved.

> **PRECEPTOR JYN**
> Exactly. You just experienced a fraction of it back in that cave.

> **ELWIN**
> I didn't mean on the space pioneers.

Elwin points to Mom at the table. She dabs her eyes with a TISSUE while DAD talks gently to their children.

> **ELWIN**
> It was hard on the people who journeyed above, but also on their families back home.

Preceptor Jyn has to think about this. It wasn't
something she considered.

> **ELWIN**
> Imagine you're one of those "happy campers"
> you keep talking about. The type of
> adventurer who thrives on being outdoors
> solving problems, collaborating with others
> in weird situations.

Preceptor Jyn is beginning to see what he means.

> **PRECEPTOR JYN**
> You would draw strength from going abroad.
> Though challenging, it'd fulfill something
> deep inside you.

Preceptor Jyn has an epiphany. She turns to Elwin.

> **PRECEPTOR JYN**
> You're sad about someone you left back on
> Kepler-62 e?

Beat. Elwin doesn't say anything at first.

> **ELWIN**
> My mom.

Elwin continues watching the simulated Mom at the
table.

> **ELWIN**
> I remember a conversation like this once.

Elwin turns his attention to the kids.

> **PRECEPTOR JYN**
> What do you think their dad's telling them
> right now?

                    ELWIN
Probably something nice, how it's only
going to be a few months.

Preceptor Jyn touches Elwin gently on the shoulder.

                 PRECEPTOR JYN
That's what many of them thought. They
must've had plans to return. To come back
someday.

Elwin lets this sink in. Then turns away.

                 PRECEPTOR JYN
Back then, people's minds were different.
They didn't think in terms of centuries or
millennia like we do today.

                    ELWIN
But this wasn't so long ago. Just 200 years
or so.

                 PRECEPTOR JYN
Yes, but back then people didn't have the
life spans you now enjoy. Rushing from
one thing to the next, many of them were
caught up in trivial things.

Elwin stares off into the distance.

                 PRECEPTOR JYN
This was still in The Decline. Back then,
people were lonely. Addicted to their
phones, they spent their days staring at
screens, updating their social profiles.

                    ELWIN
I've heard that.

> **PRECEPTOR JYN**
> For some, their world, their whole lives,
> were about getting "likes" for their posts
> or showing off what they ate for breakfast.
> As you can imagine, it was hard for them to
> give up their phones for weeks or months.

> **ELWIN**
> Doesn't seem like a big deal to me.

> **PRECEPTOR JYN**
> That's because you didn't live back then.
> This was before your time.

> **ELWIN**
> Yes, but it's still familiar.

Elwin takes one last look at the family before . . .

                                        RESUME:

INT. SPACESHIP (QUARTERS) - CONTINUOUS
Once again, Elwin and Preceptor Jyn are in his
bedroom. Elwin paces as he broods.

> **PRECEPTOR JYN**
> Something's wrong. Why did you accept this
> trip back to Earth?

> **ELWIN**
> I didn't accept it. I volunteered.

> **PRECEPTOR JYN**
> That's even weirder. Most people don't go back.
> (beat)
> I know a lot about your people's past.
> I know about science, math, many other
> things. *This* I don't understand. Why you
> came here.

Elwin can't help but laugh bitterly.

> **ELWIN**
>
> You mean there's something I know the AI teacher doesn't?

Preceptor Jyn smiles.

> **PRECEPTOR JYN**
>
> It would appear so.

> **ELWIN**
>
> Did you know my dad went to Human Spaceflight School?

> **PRECEPTOR JYN**
>
> Yes, it was in your file.

> **ELWIN**
>
> And since you know so much about human history, you'd know they didn't take everyone who applied.

> **PRECEPTOR JYN**
>
> Logistically, they couldn't. There weren't enough openings. That's why they required applications.

> **ELWIN**
>
> A lot like how I got into the Space Academy many years later.

> **PRECEPTOR JYN**
>
> Yes, of course.

Elwin goes over to the chest of drawers beside his bed. He quickly finds what he's seeking: an analog PICTURE.

He hands it to Preceptor JYN.

She sees a smiling image of MR. MATER, Elwin's dad.
He sits on the beach beside what appears to be five-
year-old Elwin and MRS. MATER, Elwin's mom.

>   **PRECEPTOR JYN**
>   What a sweet memento. It wasn't in your
>   file. You have more?

>   **ELWIN**
>   That was the only time I ever went
>   anywhere with my dad. He was paralyzed a
>   year after I was born. Car accident.

Preceptor Jyn does mental calculations.

>   **PRECEPTOR JYN**
>   But he applied for the program back when
>   you were in high school.

>   FLASHBACK TO:

INT. MATER FAMILY HOUSEHOLD - DAY
We see TEEN ELWIN (15) and his MOM and Dad sitting at
the kitchen table. (The scene is reminiscent of the
tableau we just experienced.)

>   CLOSE ON:

Dad talks to Teen Elwin as he prepares to leave for
space.

We don't hear his words, but we can see the gentleness
in Dad's face, also the way Mom comforts Teen Elwin.

>   BACK TO:

INT. SPACESHIP (QUARTERS) - CONTINUOUS
Preceptor Jyn returns the photo to Elwin.

                    ELWIN

My dad was a mess before he left. You're
looking at the only time I remember him
smiling. Space changed all that for him.

                PRECEPTOR JYN

He was in decline before he left?

Elwin nods.

                    ELWIN

Mom and I saw him a few times before he
moved to Kepler permanently. He looked
happy on the teleconnect calls.

                PRECEPTOR JYN

So that's why you followed him there. To
planet Kepler all those years later.

                    ELWIN

Yes, me and my mom. By the time we got
there, they had the science figured out
that would have allowed him to live as
long as we do now. But he was already
gone. Stroke.

Preceptor Jyn smiles compassionately.

                PRECEPTOR JYN

In microgravity, his paralysis wasn't a
hindrance. He didn't have to walk.

                    ELWIN

He could float anywhere he wanted.

                PRECEPTOR JYN

I'll ask you again: *Why* did you come back,
Elwin? Why did you volunteer for this mission?

Elwin contemplates the family photo.

                  ELWIN
      I wanted to remember . . . us.

Suddenly, an ALERT sounds as the walls turn a bright
pink.

             PRECEPTOR JYN
      We are coming up on it.

                  ELWIN
   (as if he just remembered where they were going)
      Earth.

Preceptor Jyn steps out of the way so Elwin can
approach the panoramic window: Earth in all its
splendor comes into view.

It's so beautiful Elwin has to sit down.

9

EARTH BASE ONE™: SPACE'S GRAND CENTRAL STATION

Like so many boys and girls of his generation, coauthor Tom Vice looked up at the night sky and dreamed of space. Yet unlike most people, whose dreams of the cosmos ended when they packed up their childhood models of Apollo spacecraft, Tom built a career spanning decades focused on advancing humanity's ability to explore space in LEO—and beyond.

Now, Tom brings his strategic vision to Sierra Space, leading the charge to commercialize the beyond by making LEO available to everyone—especially the private sector.

A little more background on Tom, who's also our subject matter expert for this chapter. He joined Northrop Grumman in 1986, the same year as the space shuttle *Challenger* disaster. While many in the industry believed the *Challenger* incident might spell the end for

America's future space plans, especially for the private sector, Tom didn't agree.

The tragedy underscored the challenges of humanity's inexorable advance into space—but it did not deter him. Quite the contrary, they spurred him on. Tom constantly asked himself: *How can we do this better?*

Before long, executives at Northrop Grumman realized he was finding answers and advancing critical missions. By 2010, Tom had progressed to the position of president of the company's Technical Services. He led a team of almost 20,000 employees across all 50 states and 29 foreign countries supporting government agencies, including NASA, in creating safer, more effective technical solutions across a wide swath of aerospace projects.

In 2013, he was promoted again, setting the stage for a future with Sierra Space. As president of Aerospace Systems, Tom spearheaded Northrop Grumman's pioneering work on space-based observatories, satellites, fully autonomous intelligent systems, combat aircraft, high-powered lasers, and microelectronics.

It became clear to him at this time that America was nearly ready to make the leap into the Orbital Age, the next great industrial era, in which private citizens would live and work in LEO (what we dub space workers) supported by professional astronauts. Tom eventually retired from Northrop Grumman in 2017 after 31 years of distinguished service with the company.

Even so, he knew his departure wasn't about riding off into the sunset. Instead, it was but a springboard to approach the Orbital Age with a more entrepreneurial spirit. Just before joining Sierra Space, Tom led Aerion Supersonic as its CEO. He oversaw Aerion's push to develop *boomless* supersonic aircraft with a twist: It was environmentally friendly.

Founded upon the noble mission to bring the people of Earth closer together, it energized Tom and his hardworking team. Still, he never lost his passion for space that had animated his life's work. So, when Sierra Space approached Tom about taking the helm as CEO in 2021, he leaped at the opportunity.

From his vantage point, Tom realized this company had the right technology, the right experience, and most importantly, the right people with the skills, abilities, and tenacity necessary to unleash the Orbital Age.

Coauthor Michael Ashley sat down with Tom to discuss a key component of his team's plan to bring the private sector to LEO via Earth Base One. The 121,000-square-foot campus will serve as the "Grand Central Station of space." What follows is an in-depth Q&A with Sierra Space's CEO on its significance and what the future holds for his company, its stakeholders, and most of all, humanity.

MA: Why do you compare Earth Base One to Grand Central Station? (There's a big difference between traveling by rail and traveling into low-Earth orbit, right?)

TV: No argument there. Traveling 300 miles above the Earth and orbiting our home planet at thousands of miles an hour is utterly different than riding the rails. Despite that marked difference, Earth Base One *will* play a similar role to space travel as Grand Central Terminal once played for the railroads, and, really, the transformation of America into the world's industrial superpower. But that's not all. Just as Grand Central Station transformed our lives, opening up a whole new era of opportunity and discovery, so will Earth Base One.

As you may know, Grand Central Terminal opened about 110 years ago, shifting not only people's relationship with terrestrial travel but their relationship with the broader world around them. At the time, America was in the midst of the Industrial Revolution. Grand Central unlocked its benefits to all. In essence, it removed the chaos and relative inaccessibility of travel, enabling more and more people to get where they needed to go efficiently. And consistently.

Earth Base One will play a similar role in the private sector's expansion into LEO.

Another way we like to describe Earth Base One is as the entrance ramp for the superhighway to space. Just as railroads once revolutionized travel in America, the highway system later upended how people and products move throughout the country. Still, we needed that all-important on-ramp to take advantage of the latter. We're now building *the* facility that will serve this same purpose for the Orbital Age, and it's the most exciting thing I've ever been involved with.

MA: Is Earth Base One the company's new headquarters?

TV: Earth Base One will be an *expansive* facility, but I think the concept of HQs is a bygone concept. Space isn't a regional business, so we will always have multiple locations to capitalize on the rich talent we have across our country.

I think the best way to summarize our view of Earth Base One is that it isn't intended to be a headquarters facility replacing our current locations, but rather part of a growing *network* to distribute the functions of a headquarters to wherever makes the most sense at the time—based on the mission.

MA: Will Earth Base One normalize travel to LEO?

TV: I don't think travel to space will be truly *normalized* in the near future. But it will certainly help make it more *routine*. Earth Base One opens space's potential in an unprecedented manner. We're taking off for LEO and landing (softly) on a dedicated runway. No splashdowns. No fraught, high-G reentries to Earth's atmosphere.

Also, the facility will be much more open to the public than traditional launchpads. We fully expect people to watch Dream Chasers launch from our rooftop—our team members' families, too.

Our target is to open access to space to more people—that doesn't happen overnight but, rather, is a progressive move to making it more routine. More inclusive. We're not focused on carrying the elite into space but, rather, on enabling biopharma companies, technology developers, researchers, manufacturers, and other cutting-edge enterprises to send their best and brightest into LEO to do their life's best work.

Here they will develop medicines and other marvels to benefit the human race. We're transforming how to create tomorrow's technology from today's narrow opportunities aboard existing government-run space stations like the International Space Station into a private-sector space ecosystem.

That's our vision of the Orbital Age as we've described in this book, and it all starts with Earth Base One.

MA: What does space travel "becoming more routine" mean to you?

TV: The routine part of space travel—if we can even call it that—comes into play with landings. If you think about it, launches are becoming ever more common and have been for a long time.

A SpaceX launch is thrilling to watch, but it's essentially the same thing we've seen since the 1960s, only with better cameras capturing the action. A Dream Chaser launching for LEO atop a rocket will have the same kind of look and feel of other rockets on a trajectory to achieve orbit.

But landings are a completely different beast.

Landings from space feel alien to most people. It's been more than a decade since the final space shuttle flight touched down on a runway, so landings typically occur in water or remote ground locations.

Pointedly, they happen away from the cameras. Quite often, astronauts aboard the vehicle can't even exit immediately due to hazardous materials coming off the capsule. These landings are anything *but* routine—especially as you learn more about the experience of those aboard the craft.

Most of the public doesn't know that some astronauts pass out upon landing—the g-load, the dehydration, and the re-acclimatization to Earth is just too overwhelming, and we're talking about professionals. Also, even if they don't pass out, many get quite ill.

That's not a pleasurable experience, and the human aspect of space travel is critical to the Orbital Age. Here's an example: The person best suited to discover an amazing medical breakthrough in LEO is not necessarily the one best suited to handle such a difficult trek home.

On the other hand, runway landings are routine for anyone who has flown in an airplane before. Sure, there may be some white-knuckling on descent, but you don't typically see half your passengers passing out or growing ill.

Space travel becomes more routine when astronauts and space workers can enjoy a relatively smooth landing of perhaps 1.5Gs on a runway. Within 15 minutes after landing, they'll exit the Dream Chaser to find their family waiting, along with hot showers and fresh food available in Earth Base One a short walk away.

That's what routine spaceflight will look like in the very near future.

MA: Can you describe what the Earth Base One experience will be like for companies who want to go to space?

TV: Bringing the private sector off-world is a *partnership*—we can provide the platform for companies, inventors, and researchers to go do their best work. Naturally, it'll entail a different type of relationship. It all starts when a private-sector organization understands their business can benefit from LEO's unique properties.

This might be a biopharma company tackling the deadliest forms of cancer, an outfit developing new materials to protect first responders, or innovators in industries we haven't even thought of yet. Once the first step has been taken of establishing interest, Earth Base One will play an important role.

Private-sector partners can expect to become frequent visitors to Earth Base One. They will be on hand long before their employees ever live and work in space. The process will begin with determining their mission's scope and what sort of equipment will be needed to accomplish their R&D or production tasks in LEO.

A key feature of Earth Base One's production capabilities is the fact that we will customize LIFE Habitats and, to some extent, Dream Chaser configurations to suit our partners. The private sector has always balked at NASA's "one size fits all" approach. They rightly expect facilities to suit their needs and help them achieve success, whether it's a new production facility down the street, or one 300 miles up in LEO.

During this initial phase, we'll answer all of the difficult engineering questions, like how specialized equipment can be powered while leaving enough juice to keep the lights on. As this nascent work continues, we'll review our partners' candidates to become space workers.

Aspirants will go through an evaluation period to determine if they are a good mission fit. Although the physical requirements will be different than for professional astronauts, space workers will still require several months of training at Earth Base One to qualify to live and work in space.

After both the space workers' training and all technical tasks have been completed, our partners will be ready to begin their space operations.

We anticipate partner organizations will keep staff at Earth Base One to act as a liaison throughout the process. This allows for convenient access to mission control if there are any questions, along with providing the most secure communications with their space workers. Also, when Dream Chaser flights return with the results of experiments or goods produced in microgravity, the liaison will be on hand to take control of cargo within an hour of landing.

MA: Zooming in on Sierra Space technology, what impact will Earth Base One have on the Dream Chaser spaceplane?

TV: Earth Base One is the first of many operational centers for the Dream Chaser spaceplane. Dream Chaser, *Tenacity*™, and our second vehicle, *Reverence*, are being built at our Colorado facility. Our Dream Factory in Colorado will continue to be our advanced spaceplane development center.

Earth Base One is an integration, test, launch, and mission operations hub.

Every aspect of Dream Chaser operations will be managed here. Before a mission, the spaceplane will be configured with the particular profile with all cargo, crew, and passengers accounted for. The flight will be controlled from beginning to end by one of several mission control facilities; the primary centers are in our Colorado campus, with a backup center at Earth Base One.

Then, after landing at Earth Base One, the cargo will be appropriately handled to get it safely to its final destination. Any repairs or modifications to the spaceplane will be completed right at the base. The ability to quickly turn around a Dream Chaser after landing and have it ready to go again is a major step forward to achieving the type of flexibility the private sector requires from its partners.

MA: What impact will Earth Base One have on the LIFE Habitat and other technologies Sierra Space is developing?

TV: LIFE Habitats will be built at Earth Base One. The first wave of construction at the base is concentrated on our production facility, because the demand for systems like the LIFE Habitat, the backbone of the Orbital Age, outstrips our current manufacturing facilities.

So, LIFE modules will be built at Earth Base One, then customized or configured to suit the specific needs of a particular space station plan. LIFE Habitats will launch from the LLF, just a matter of yards from where they were constructed, then be inflated in space, ready to dock with other modules and, eventually, the first Dream Chaser carrying station crew.

Although the production emphasis will be on Dream Chaser spaceplanes and LIFE Habitats, we'll also have room to spin up production on other products and space-enabling technologies. In

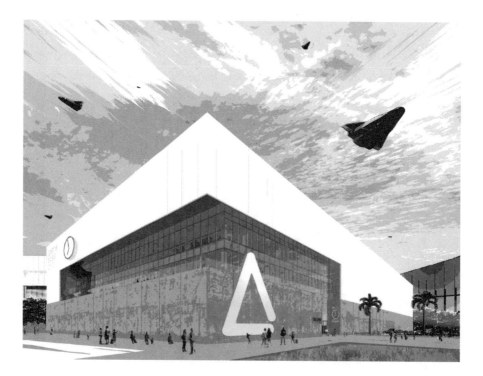

short, two key advantages of production at Earth Base One are that we'll have the size to build whatever we need, and access to the best pool of high-tech manufacturing talent to turn designs into reality.

MA: Let's focus on the name. Does the moniker Earth Base One tell us anything about your future plans?

TV: Yes, it does! Earth Base One is the Grand Central Station of space, but the latter wasn't very useful if you were traveling from Chicago to Philadelphia. For the same reason, we envision a network of Earth Bases, of which this is the first.

We will create this in the US at the same time as we expand internationally—because space is for everyone, including people on the other side of the world.

Even so, we feel America must maintain its leadership position in space, which we will achieve by keeping our flagship Earth Base in the United States. Eventually, Earth Base One will be the largest of dozens of Earth Bases, each dedicated to making space travel as routine as possible for their area.

Even as Earth Base One enters operational duty, we'll be evaluating locations for future bases. As we enter the Orbital Age, launches to LEO will become less synonymous with one US location. (We already have SpaceX frequently launching to space from Texas, and that trend will continue to expand.)

The Earth Base network will likewise grow as part of this trend in developing more locations that meet specific criteria. First and foremost, we will evaluate the geographic area to determine if an Earth Base makes sense based on the mix of regional industries. (If companies exist in a given area that would greatly benefit from expanding into LEO and are ready to go, a more immediate need for a base is warranted.)

For example, I anticipate great interest to come from North Carolina's Research Triangle. Is it reasonable for tomorrow's bio-tech startups and other adjacent high-tech pioneers located there to carry their very delicate cargo all the way to, say, Florida's Space Coast? Or is that market better served by a nearby Earth Base? We'll be evaluating this question along with others.

Besides the desire to locate Earth Bases to best serve partners, there are many other factors to consider. Here's one: Can we ade-quately staff a facility with a mix of transplants and local talent?

Likewise, is there an existing runway long enough for Dream Chaser landings, or can we easily construct or expand one? Here's another concern: Will the base also facilitate launches, or will it be strictly for landings? These are just a few of the variables that will go into the Earth Base selection process.

MA: Do you envision producing similar bases off-world in the future?

TV: Creating bases beyond Earth will be a natural progression as the Orbital Age reaches maturity. Our philosophy is that we are building the infrastructure to make space available to everyone.

To achieve our goal, we must methodically approach each step to ensure we have solid footing for the challenges we shall face. Right now, we're building on a foundation of decades of work in space by intrepid pioneers. The network of proposed Earth Bases, along with received insights, a refinement of processes, and a vast trove of mission data to come, will provide the basis for creating permanent facilities beyond Earth.

Once we have a formative network of Earth Bases, the next log-ical step is to build Moon Base One. Although some in the space industry are quite fixated on Mars as the next major destination, we believe humanity won't successfully achieve ongoing missions to the Red Planet. Yet.

Not until the moon has one or more permanent bases in operation.

You could write a whole book on just the topic of how Moon Base One will help our collective efforts to reach Mars. For now, let me say, a permanent off-planet base will especially help us to better discern the *psychological* challenges of being far from Earth for an extended period.

Tomorrow's astronauts living and working on the moon will be able to create procedures for emergencies and other safety challenges. They'll also learn more about medical care and triage on these long missions. The last thing anyone should want is for a Mars voyage to encounter suit breaches, toothaches among personnel, or any of the million other things that can go wrong in space for the first time—so you run constant exercises on the moon first.

We look at it this way. Today we're focused on making travel to LEO routine. Once accomplished, we will use our experience and findings from this work to, in turn, make traveling to the moon routine by constructing Moon Base One.

Once that's *also* routine, we'll talk about Mars Base One. Such a rational, methodical approach is how space exploration can be transformed into successful space commercialization to enhance life back on Earth.

MA: Would it be reasonable to compare Earth Base One to an airport when describing it in simplistic terms, perhaps to a child?

TV: Yes, that is a reasonable comparison on a few levels. Airports are the essential component enabling air travel for the public. They provide an end-to-end solution for travelers.

You don't have to show up at O'Hare International Airport with your own pilot, navigation charts, or a spare fan blade in case an engine needs servicing. All of those aspects are handled by staff onsite, so that private citizens can focus on where, how, and why they are traveling.

Earth Base One will provide a similar experience for a very different destination. Companies in the private sector, whether they are a biotech powerhouse or an energy startup, have little to no interest in developing core competencies to travel to space safely and routinely.

Now, they won't have to.

By utilizing Earth Base One, the Grand Central Station of space, they'll find the vehicle to get them to orbit and back, the Dream Chaser spaceplane. They'll find their destination in LEO, the LIFE Habitats that will combine into future space stations. They'll find the tech underpinning all that will make their work possible with our applications business. They'll also even find the ability to power their projects with our next generation of solar panels and other innovations.

Today, you go to an airport because they handle everything related to air travel. Earth Base One will do the same thing for destinations 300 miles above us.

MA: Throughout this book we've indicated a big shift is coming to space travel. Can you explain it and the role Earth Base One will play in the transition?

TV: The shift in space travel is the central component of the Orbital Age. Up to this point, space travel has been about exploration and experimentation. We've made many discoveries in space, whether they were aboard the ISS, or even thanks to marvelous

telescopes for answering the mysteries of the universe by peering back through time.

That is all important work, but now we should change focus from exploring to executing in space. This means the production of medicines, research on exotic materials, and developments in energy and other industries to immediately improve life on Earth.

At the same time, we must open the door to space for more people.

If you consider the history of human space travel, an astonishingly small number of humans have left our home planet. Far fewer than a thousand people have ever reached low-Earth orbit, out of billions upon billions of humans.

Earth Base One is a critical component to changing that. By

making travel to LEO more routine, more people will be traveling to space. We won't be in a position where millions will be in orbit anytime soon, but we can surpass the total number of astronauts thanks to this facility and those like it in the future.

MA: What impact will the Orbital Age and Earth Base One have on the American economy?

TV: Space workers—private-sector professionals living and working in LEO—will produce a staggering explosion of growth. Consider how transformative the Internet Age has been to the economy. There are highly paid careers in diverse fields that didn't exist 20 years ago, like search engine optimization [SEO].

The Internet Age allowed people and companies to move faster, learn more quickly, and be progressive. The Orbital Age will have a similar impact.

Space's potential is wrapped up in how we can do things in microgravity's unique environment—something impossible or difficult to achieve on Earth. We can not only discover medical advances faster in space, but we can manufacture medicine in new ways thanks to lack of gravity. Whole new industries will emerge, like manufacturing with hazardous materials to reduce the risk to Earth's environment, not to mention orbiting data centers that are more efficient to produce off-planet.

MA: As we move quickly into the Orbital Age, how do we ensure our technology does not outstrip our ethics?

TV: This is a very important topic—how do we go to space and not damage it much like we have harmed our oceans and other natural resources? Sierra Space believes we must learn from humanity's mistakes, so the same thing doesn't occur in LEO.

Already, we have a space debris problem, with certain orbital altitudes essentially turning into space junkyards. To prevent further damage, we must act as space stewards. This comes with the responsibility to look after the orbital environment much more conscientiously. Easier said than done, I know, especially because there isn't much holding others to this view of care.

The only legal agreement in place with any teeth is NASA's non-binding Artemis Accords. These are broad principles describing how NASA believes humans should use space. This must evolve into a more practical and ethical code of conduct to keep pace with the evolution of space.

Humanity has an opportunity at this moment to start with as close to a clean slate as modern humans can imagine. This time, we can get it right from the start with a new moral code to safeguard space for future generations. This becomes both more complicated and more important as we ratchet up our commercialization efforts.

It matters, though. The importance of taking care of the environment, even if it is a cold void vast in size and scope, is critical to tomorrow's undertakings. Just like our Blue Dot, we have a collective responsibility to look after space.

MA: So, what would a space code of conduct look like?

TV: The focus will be on getting agreement from the entire space community—whether nation-states or private companies—on how this novel environment should be treated and maintained. This will undoubtedly look and feel different concerning various topics, especially so that it remains fair and beneficial for all parties.

First, let's use the example of space debris. As mentioned, it's one of the existing problems in orbit that would hinder the Orbital Age if it worsened.

A code of conduct around space junk would be simple and pragmatic. If you make a mess in space, you clean it up or pay to have it cleaned up. There's no reason to have dormant equipment and other debris hurtling in orbit around Earth. Ensuring it's managed is a commonsense component of what I'm suggesting.

Naturally, there are thornier issues that international consensus will be required to solve, such as to what extent terrestrial rules apply in space. For example, will there be such a thing as "national territory" as we now observe it with our oceans, or will it all be "international waters"?

That's the sort of question we'll need to answer sooner rather than later.

MA: Can humanity's leap into space protect those of us back on Earth as well?

TV: There will always be a national security element to space. In fact, a global security aspect too. Having more people in LEO—and eventually beyond—will provide a better set of "eyes and ears" for potential threats than we've ever possessed.

Space workers and professional astronauts in the Orbital Age will have a prime observation spot for not only watching Earth, but also looking outward. I can imagine an object, like a meteor, that might cause a major natural disaster if it hits the Earth being diverted using technology and skills gained through the operation of Dream Chaser spaceplanes and LIFE Habitat space stations.

MA: Last question: What does space travel look like 100 years from now?

TV: In 100 years, space travel will not be as common as air travel is today, but it will be well on its way.

Being a space worker for four or six months at a time will be considered a prestige assignment, a sign you've "made it" as a researcher or engineer, but not uncommon by any means.

People will marvel at the concept of NASA once having an Astronaut Hall of Fame. Space travel will be common enough that they'll have trouble imagining a time when only a handful of intrepid explorers first left Earth's confines. It will be a similar feeling to when a person on a flight from St. Louis to Oregon imagines being part of the Lewis and Clark Expedition making that inaugural voyage.

Also, people who would never qualify to serve as professional astronauts will safely journey to space, enjoying a freedom of mobility unheard of on Earth.

Meanwhile, illnesses that remain death sentences today will be curable thanks to medicine developed and manufactured in LEO, and surgery in microgravity will be a growth industry. There will also be no organ transplant lists because replacement organs will be grown aboard orbiting labs.

Most importantly, the dreams of future generations will be more readily achieved due to countless advances in industries that could only occur *because* the private sector could send the right experts to live and work in space. You can think about it like this. Space is the new platform for business. Much like the internet presented an innovative hub for new companies to emerge in traditional industries, including commerce, hospitality, and transportation, our Sierra Space platform will power the space economy.

Based on this, the possibilities are endless. We will even enable humanity to start new civilizations in space beyond our own. That's because we envision a bright future where humanity lives and works in space, on moons, and on distant planets. Here, future people will marry, children will be born, families will be raised, businesses will be built, and most of all—humankind will thrive.

The first step is to build a platform in space to enable such discoveries. That's why Earth Base One—space's Grand Central Station—is so needed. In a word? Space travel will be better. And so will life here on Earth because of it.

10

THE GREAT CHOICE

Time: the not-so-distant future.

STORY #1: THE PERMAVERSE

Like most 22-year-olds, Alex Denehy had no use for dressing up. He never wore a suit and tie a day in his life. ("Nope. Don't even own one.") He stayed in his pajamas all day. Or at least, his version of nightwear: sweats, T-shirt, wool socks. The last item was for comfort. He lives in Whitefish, Montana, where you know . . . it gets cold—even if the only time you glimpse the outside world is to yank in the latest drone-delivered Amazon package.

It's been said that iGen was the first generation to grow up with the internet in their back pocket. Born between 1995 and 2012, the oldest were on the cusp of teenhood when the late Steve Jobs introduced the iPhone in 2007. Early adopters, their adolescences dovetailed with the rise of web 2.0. Snapchat, YouTube, Insta, and Twitter were like *Friends*' Central Perk to them: (digital) hubs—okay, homes away from home for them. *Hangouts*. Here, they checked in with friends, learned gossip, shared selfies.

It's where they grew up.

Immersive as that experience was, technical limits still constrained it. In the same way Gen Xers viewed once cutting-edge innovations like the steam engine and telegrams as woefully outmoded, Gen Alpha (first born in 2012) saw touchscreens as veritable relics. ("A *two-dimensional* internet experience? We talkin' about the Stone Age or what?")

Alex and his generation's adolescence coincided with the COVID lockdowns, widespread teleconferencing adoption, and, of course, remote instruction. To him, there was never existence without the web. That's why he embraced the *Permaverse* rollout with open arms.

Forget about being offline. Ever. He was more comfortable *in* the internet than outside it.

Though technically, Alex still lived in his parents' basement, he owned a mansion. *In the cloud.* Composed of pixels and clever coding, it provided labyrinthine hallways, vast bedrooms, an Olympic-sized pool, and a charming helipad.

In real life (IRL), of course, Alex could never hope to buy a starter home on his meager salary—that is, donations he racked up from fellow Twitchers who swooned watching him dominate first-person shooters. Yet, ensconced in his three-dimensional digital realm, he lived like a king. Donning his haptic suit, he could even enjoy tactile experiences like feeling the wind in his hair whenever he rode his *Permaverse* Harley.

For Alex and others like him, Earth and all its messy physicality (icy Montana winters, the endless risk of diseases like COVID, insects, etc.) were turnoffs. They had little use for it. A *Permavizen*,

Alex would exist 100 percent online if it weren't for pesky bodily needs like urination and eating.

The last thing on Earth he'd want to do is wallow in what's now derisively called the . . .

STORY #2: TERRAVERSE

Decades older than Alex, Jamie Elstair can recall a time before parents hosted children's birthdays in cyberspace, an era when people physically commuted to work and kids took the bus to school.

She appreciates the convenience of going to the perma-post office but would still rather visit in person. Like most of her peers, Jamie switched careers a dozen times in her life. A quick learner, she enjoyed the thrill of acquiring new skills like data sleuthing and auditing algorithm biases.

What didn't she like?

Work's drudgery. It sapped her energy, killing her spirit. Learning was enjoyable. So was making new officemates and attending conventions, even holographic ones. But Jamie was someone who needed purpose. She wanted what she did, day in and day out, to possess meaning. Her nightmare was growing old and feeling like the bulk of her existence was just . . . blah.

That's why she dropped out of the workforce during the Great Automation Resignation. The 2020 pandemic was the first big disruptor transforming how people viewed work, leading to mass exits of employees. But the effects weren't enduring. In time, inflation and a massive recession forced many of the same job defectors back to the grind. That is, until the economy stabilized within the decade and work-life returned to a new normal.

That was *also* temporary.

At the same time workers were just getting back on their feet, the rug came out from under them. Sweeping advances in AI displaced wide swathes of the workforce. And not just blue-collar workers like truck drivers or retail clerks. Vast numbers of supposedly safe white-collar jobs vanished: lawyers, doctors, CPAs. While specialties within these fields remained viable (think: brain surgeons, IP patent attorneys focusing on robotics contracts), many generalists succumbed to the vagaries of wholesale automation.

To stave off societal unrest, the government issued universal basic income (UBI) to those affected by the disruption. Unintended—but predictable—consequences ensued. *Why should I work if Uncle Sam's handing out cash?* went the thinking, as huge blocks of professionals abandoned the workforce in protest.

Before long, a second, more massive wave of mass resignations rippled through the nation. Then the globe.

Fallout from the white-collar walkout spilled over to other sectors:

- Burned-out nurses called it quits.

- Frazzled teachers stopped showing up to school.

- Overworked first responders (EMTs, firefighters, and police officers) refused to put themselves on the line anymore.

Before long, daily physical life became bleaker and more challenging. It pushed ever more people, especially younger ones, to the *Permaverse*. Virtual reality become more appealing than physical reality. The losers? Folks Jamie's age and older who still live and work IRL.

Then humanity's savior came in an unlikely form . . .

STORY #3: SPATIAVERSE

The first film Angel Escobar can remember seeing was the original *Star Wars* (1977). Decades old and sporting outdated special effects, it still lit a fire in his belly to go to space. While his peers were busy choreographing dance numbers on TikTok Perma and designing AI pets, he gorged on literature. Science fiction stories, to be exact. He all but inhaled the novels of Robert Heinlein, Philip K. Dick, Ray Bradbury, and Isaac Asimov.

Forget about living in his folks' basement or accepting one of the few terrestrial jobs holding any allure; Angel knew his destiny lay in the stars. In high school, he took scuba training to prepare to spacewalk. He gorged on online courses exploring astronomy and physics.

It was Angel's dream to leave Earth, to become a true explorer in the mold of Lewis and Clark or Marco Polo.

Restlessness built within him.

His parents noticed it first. After securing the top GPA in his junior class, he researched the highest GPA historically at his high school. Besting that number became his new goal. Once he achieved that, he changed how he competed altogether.

The race went within. Grades lost all their meaning. So did test scores. And academic awards. Instead, he made new goals for himself:

1. Read one new book a week *day*

2. ~~Practice~~ *Learn* a new language

3. ~~Listen to a~~ Memorize a great speech

Despite the *Permaverse*'s great promise, Angel's peers couldn't relate to his existential zest, his desire for achievement. They were bored online, unhappy with their lives. Many couldn't put their finger on why. It was as if they were suffering from the malaise kids get the day after Christmas. With all the presents opened and all the anticipation gone, what's there to do now?

Angel never suffered such feelings.

Life was the great adventure to him. Every moment was preparation toward his goal of space exploration. That's why he graduated high school early. Instead of rushing off to college, he took his own gap year. In his case, the gap was interning at Acme Space Company.

It was more competitive to get into than anything he ever tried before, but Angel liked the challenge. It imbued his struggle with meaning. To borrow an iconic phrase, it meant Angel had the "right stuff."

After months of testing and exams, Angel learned he had been accepted for a paid internship. Should he do well, he would advance to a team destined to colonize the moon.

Learning the news, Angel cried with happiness. His dream had come true. More importantly, Angel and other pioneers like him had just opened the newest, most exciting chapter in the continuing story of our human race.

THESE THREE STORIES ARE fiction but based on reality. The second tale describes a real and growing disaffection toward work, a trend dubbed the Great Resignation or the Great Reshuffle. Before we explore this topic, let us also note that should current workplace trends continue, we can expect to see more young people like Alex in Story #1 opt out of the (physical) economy, not to mention check out of IRL.

Now, let's define terms. *Forbes* contributor Christine Comaford, reflecting on the abundance of uncertainty, fear, and economic disruption caused by the pandemic, explains what precipitated the Great Resignation this way in a January 14, 2022, piece:

> Workers began to reflect on their relationship with work. And as they did, they realized that certain aspects of work just didn't work for them anymore. . . the daily grind suddenly seemed excessive. The call to a higher quality of life—as we considered our mortality—was deafening.[1]

Laura Conover, president of Conover Consulting and an expert on corporate culture, suggests there are deeper issues at play. Beyond workplace annoyances, many employees had come to experience widespread disillusionment with their jobs. The pandemic was just the catalyst that sent them running for the door. "For the fact is, no amount of money can compensate for a working environment offering only frustration, dehumanization, abuse, confusion, belittlement, anonymity, and/or soul-sucking boredom," she wrote for LinkedIn in an article entitled "How to Become an Employer of Choice in 2021."[2]

This chapter is all about how to do just what Conover is suggesting: become an employer of choice—and stave off the next mass employment exodus. But first, if you will recall, Jamie in Story #2 suffered similar feelings of the drudgery Conover describes. Working for working's sake didn't fulfill her. She succumbed to burnout. She wasn't alone in our tale. Our fictional white-collar walkout spilled over to other sectors, including nurses, teachers, police officers.

This is *already* happening. Consider these alarming (real-life) statistics:

Nurses Quitting

"Nearly 1 in 5 health care workers has quit their job during the pandemic."[3]

Teachers Quitting

"In Florida, teacher vacancies this year increased by more than 67 percent, compared with August 2020, and a 38.7 percent increase from August 2019."[4]

Police Quitting

"The Las Vegas Metropolitan Police Department, for example, saw retirements increase by 37 percent in 2020. . . . Police shortages are so severe in some cities that they're recruiting using billboard signs."[5]

Growing up in the shadow of the pandemic and the Great Quit, it's little wonder that Alex in Story #1 never bothered with a traditional job in the first place. Gen Z (those born from 1996 to 2012) is already being called the Great Resignation Generation for its propensity to job-hop or outright refusal to work.

As Hillary Hoffower wrote for *Business Insider* in February 2022: "Gen Z is ready to leave their jobs in the dust. Sixty-five percent of the generation plans to quit their job this year . . ."[6] If current trends continue, we can imagine a future in which more young people take Alex's approach, dropping out of both the economy and real life in favor of a full-time online existence.

Unprecedented might be a good word to describe this situation, not to mention a coming time encompassing all three above stories. In many ways, this future will be without parallel. Never will the world have seen so many changes, both technological and societal—and never so fast. Even the last few decades, in which the Internet utterly redefined daily life and commerce, fail to compare to the magnitude of the impending sea change.

Before we know it, people will soon be able to experience the three ways of being (sometimes simultaneously) that we have coined *The Permaverse*, *The Terraverse*, and *The Spatiaverse*.

The Permaverse:

Our story uses this fictional moniker based on the real Metaverse. "The Metaverse is a collective virtual open space, created by the convergence of virtually enhanced physical and digital reality," explains Ashutosh Gupta for Gartner. "It is physically persistent and provides enhanced immersive experiences. Activities that take place in isolated environments (buying digital land and constructing virtual homes, participating in a virtual social experience, etc.) will eventually take place in the Metaverse."[7]

The Terraverse:

This is the IRL reality we inhabit now, the physical plane of existence that's sustained humankind since the dawn of time, the sole home we have ever known. It has only been (very) recently that new technologies like the web have begun to supplant Earth's physicality. No doubt you've experienced this: texting someone in your office rather than asking a question in person. Interacting with clients 100 percent online whom you've never met in person. Digital screens now mediate much of reality, especially for people who live in cities. As Allison Goldberg explained for *Forbes*: "We live in a world where we can do almost anything with a few taps of a screen. . . . It is now socially acceptable to talk to seven other people while in real-life conversation with one."[8]

The Spatiaverse:

Emerging tech, such as the world's first business park in space we've covered so often in this book, present the latest mode of being. Living and working off-world will soon no longer be the stuff of science fiction. In the same way so-called digital natives (Gen Z) grew up with the internet in their back pocket and have no concept of life before it, something analogous will soon happen, beginning in low-Earth orbit. For future generations, their existential baseline will be a reality where space existence is *commonplace*, much like today's toddlers use phones and tablets to do things online that would have seemed incredible 100 years ago. Or our favorite quote—as futurist/science fiction author Arthur C. Clarke put the incongruity of ever more complex innovation so well: "Any sufficiently advanced technology is indistinguishable from magic."

TAKING A STEP BACK, it turns out there *is* some precedent for this moment. The 1800s in the fledgling United States of America offers a powerful corollary. An era alive with exciting innovations and immense possibilities, it straddled the old world and the new. Harry L. Katz and the Library of Congress captured this unique period in the book *Mark Twain's America: A Celebration in Words and Images*:

In 1835, the year Sam Clemens [Mark Twain] was born, rivers powered the nation's commerce, transportation, and communications, serving as highways for freight and passengers. . . .

Steamboats, barges, rafts, and riverboats delivered news and goods across the nation, linking rural homesteads and city parlors. For Clemens and many of his compatriots, rivers also represented the allure of exploration and escape. Unknown vistas and untold opportunities lay around the next bend. During the 1820s, 1830s, and 1840s, American artists and writers in the East found inspiration in the remote regions of the Hudson River, producing landscapes, poems, and novels portraying the Hudson River Valley as wild, picturesque, mysterious, and a sublime metaphor for the human spirit. . . .

Transcendentalist writers, led by Ralph Waldo Emerson and Henry David Thoreau, further extolled wilderness as an expression of free will and a source of enlightenment. James Fenimore Cooper's popular novels conferred natural nobility on the backwoodsmen, trappers, soldiers, settlers, and Native Americans who lived in the pristine forests. This early generation of artists and writers treated wilderness as an abstract concept, an escape valve for the soul.[9]

In many ways, this period fostered a kind of American Renaissance, an unprecedented time of profound innovation, imagination, and growth. Let us count but a few of the many ways:

1803: The third US president, Thomas Jefferson, enacted the Louisiana Purchase from France, acquiring nearly 850,000 square miles of territory, doubling the nation's size.

1814: The same year engineer George Stephenson designed the first steam locomotive, the first instance of plastic surgery was performed.

1850: Joel Houghton achieved a patent for the first dishwasher. Constructed of wood, it employed a hand-turn wheel to splash water on your dishes.

1851: Emanuel Leutze painted *Washington Crossing the Delaware*. Mythically symbolic, it aided in the construction of a national identity, celebrating its founding. This same year, Herman Melville published *Moby-Dick*, one of the most iconic novels of the American canon.

1855: Walt Whitman published the poetry book *Leaves of Grass*, ushering in a new era of American literature and artistic expression. American transcendentalism is one such category, producing intellectual luminaries like Emerson and Thoreau.

1869: The transcontinental railroad was completed, unifying the United States from coast to coast. Featuring 2,000 miles of track, it reduced journeys from months by wagon or boat to mere days. Moreover, it enabled the young US to become a leader on the world stage with its highly productive, industrialized economy.

1876: Alexander Graham Bell patented the telephone.

1877: Thomas Edison invented the phonograph.

1890: American philosopher William James published his 1,200-page masterwork, *The Principles of Psychology*, leading to modern psychology and (now commonplace) concepts such as "the stream of thought."

Reflecting on these historic developments (along with many more unmentioned) permits us a solution to the so-called Great Resignation. As we have observed, lack of purpose, meaning, and joy (often reflected in feelings of burnout) are producing mass disaffection, forcing employees to question daily existence, especially work, the principal activity occupying much of our time.

Ashley Grice, writing for *Fortune*, tapped into this widespread phenomenon in September 2021:

> With fresh perspective, we search for a focal point in our lives—something that consistently draws us in, and from which everything else can flow out. When we struggle to find this, we feel lost and often with an unclear vision of an ideal future.[10]

Extrapolating from Grice and Conover's finger-on-the-pulse diagnosis of the zeitgeist, it's clear we must inspire future workers in novel ways. Otherwise, most companies won't just not be employers of choice—they won't be employers at all. Should this happen, we can scarcely imagine the deleterious consequences on the economy, the world, and our way of life.

The *Spatiaverse*, as we term it, offers a solution to the future-of-work crisis. To borrow a phrase from Katz, more than any other technological feat, even the internet, it offers "unknown vistas and untold opportunities" ripe for novel inventions, works of art, even new philosophies. Beyond dramatizing its allure in several of the chapters of this book, let's close by discussing yet another way space explorers could revolutionize life on Earth. And beyond.

Our example portends positive things to come, should we only reframe today's Great Resignation as tomorrow's Great Choice. That choice, of course, is to select a career or business in our nascent commercial space economy.

In our Orbital Age.

HOW FUTURE SPACE EXPLORERS MIGHT TRANSFORM CIVILIZATION

In 1964, Russian astrophysicist Nikolai Kardashev created the so-called Kardashev Scale, a method for measuring a civilization's level of advancement based on the amount of energy it is capable of summoning. At the time, he sought evidence of extraterrestrial life. His assumption was that as planetary civilizations grow larger or more complex, their energy demands would necessarily increase.

What follows is a discussion of the various civilization types. Note: Before delving into their specific features, it's essential to realize Earth's *current* limitations. Our Pale Blue Dot—with all its incredible innovations (supersonic aircraft, smartphones, quantum computers, even the internet)—doesn't make the list. We are considered a Type 0 civilization, as we still principally harness our energy needs from deceased animals and plants.

Type I Designation: Planetary-Level

A civilization of this caliber would be capable of leveraging *all* the available energy on its parent planet. "Being able to harness all Earth's energy would also mean that we could have control over all natural forces. Human beings could control volcanoes, the weather, and even earthquakes! (At least, that is the idea.) These kinds of feats are hard to believe but compared to the advances that may still be to come, these are just basic and primitive levels of control (it's absolutely nothing compared to the capabilities of societies with higher rankings)," explains Jolene Creighton for *Futurism*.[11] Furthermore, experts such as physicist Michio Kaku believe we are still 100 to 200 years away from such a development. It would require increasing our current energy production by a whopping 100,000 times what it is now.

Type II Designation: Star-Level

A full thousand years in the making from Earth's current state, the next stage involves harnessing the power of an *entire star*. Not just converting starlight into energy, but exploiting the star in its totality. "And in order to absorb this huge amount of energy, they [would need] a super tool called the Dyson structure," explains scienceinfo.net. "For ease of understanding, the Dyson structure is a dense system of satellites that envelops the star and draws its energy."[12]

This kind of power dwarfs today's capabilities to an almost unimaginable scale. (But we'll try!) Fans of the film *Don't Look Up* (2021) will recall that even the brightest minds on Earth couldn't figure out how to escape an impending asteroid from colliding into Earth, destroying all life. A Type II civilization wouldn't suffer the same fate. Even if a moon-sized object penetrated our solar system en route to decimate us dinosaur 2.0–style, we'd still be able to thwart it based on our profound powers. Our preferred disposal method? Vaporization. We could also move Jupiter into its path, sparing Earth from total annihilation.

Type III Designation: Galactic-Level

Being able to thwart planetary extinction from intruding heavenly bodies is no small deed. But it's just a warm-up for what's possible next. Imagine being able to control an entire *galaxy*. Some experts think the only way to ascend to a Type III civilization is to harness energy via leveraging supermassive black holes—which have been proven to exist at the center of large galaxies. They also predict that it may take us hundreds of thousands, if not millions, of years to reach this advanced civilization level.

No matter how long it would take to get there, it's mind-bending to contemplate the possibilities such power affords. Fans of Douglas Adams's sci-fi novel *The Hitchhiker's Guide to the Galaxy* will be familiar with one such fantastical capability: planet creation. "And thus, were created the conditions for a staggering new form of specialist industry: custom-made luxury planet building. The home of this industry was the planet Magrathea, where hyperspatial engineers sucked matter through white holes in space to form it into dream planets—gold planets, platinum planets, soft rubber planets with lots of earthquakes—all lovingly made to meet the exacting standards that the Galaxy's richest men naturally came to expect."[13]

Yes. That's the kind of power we are talking about.

Ultimately, Kardashev didn't believe civilizations could transcend our limitations. To him, they represented the apotheosis of ingenuity, human or otherwise. Yet other theoreticians think differently. There are those who suggest greater vistas exist.

We will briefly describe their conjectures below.

Type IV Designation: Universe-Level

This civilization would be able to employ the energy content of the entire universe. Enjoyers of such power could go anywhere at any time across all galaxies, even dwelling inside supermassive black holes, subverting all our conventional understandings of physics.

Type V Designation: Multiverse-Level

The multiverse theory suggests there isn't just one universe but many, a vast cacophony of universes, each teeming with their own galaxies, solar systems, planets, and species. If a multiverse does exist, this civilization wouldn't just be able to manipulate their own universe but others too, enabling them with nearly godlike powers beyond our wildest understanding.

OKAY.

This discussion is (literally) far out, the kind of scientific inquiry most laypeople rarely indulge in. But it has a practical purpose. It's meant to bend your mind—to stretch the limits of what you deem possible. The opposite of pedestrian, it suggests other ways of looking at reality you may not have considered.

Returning to our challenge: engaging tomorrow's workforce. It's our contention that workers who quit because of the Great Resignation did so because they lacked any one of three things (or all of them):

- Meaning

- Purpose

- Joy

To be fair, workplace toxicity is a big reason people leave, along with too little pay, but we are going to leave these factors out, as they vary by individual or organization. Instead, we wish to propose a scenario whereby tomorrow's space explorers can satisfy all three requirements, leading to their Great Choice—to stay invested in the economy, the *commercial space economy*.

Picture this. It's been several years since Angel Escobar was first accepted for his internship with Acme Space Company. A fast, eager learner, Acme hired Angel to work full-time within months of starting his internship.

Inspired by Kardashev's paradigm, Acme had years ago made the bold choice to be the company that terraformed Mars, making it hospitable for humans to live there, ultimately transforming Earth into a Type I civilization.

In effect, Acme wished to achieve SpaceX CEO Elon Musk's dream of becoming a multiplanetary species. In his own words: "I think fundamentally the future is vastly more exciting and interesting if we're a spacefaring civilization and a multiplanet species than if we're not. You want to be inspired by things. You want to wake up in the morning and think the future is going to be great. And that's what being a spacefaring civilization is all about."[14]

Of course, achieving the dream of colonizing Mars predates Musk. Many science fiction authors, including Ray Bradbury (*The Martian Chronicles*) and C. S. Lewis (*Out of the Silent Planet*), seeded the idea, actually influencing him and others to dream big with their space companies.

Though Earth was nowhere close to becoming a Type I planet when Angel started at Acme, from day one on the job he knew that was the mission—by first converting Mars into a prototypical sandbox. "The goal of our collaboration is to return the benefits to Earth," said Caroline Masters, Acme's CEO, at the first teleconference Angel attended. "Each one of you helps us to achieve this purpose every day through your tenacity. Your creativity."

Masters's words weren't mere lip service, and Angel took them to heart.

Bold, they articulated a vision that all 3,000 of his fellow workers could not just get behind but strive daily to pull off. While most of Angel's peers back on Earth had checked into *The Permaverse*, he was busily engaged in work that gave his life meaning, activities that filled him with wonder and purpose.

That was good because the work wasn't easy.

Every day he awoke in microgravity, zipped into a sleeping bag strapped to the bulkhead of the modular space station where he was assigned. Separated from all his friends and family, he suffered difficult bouts of loneliness exacerbated by challenging physical demands like keeping up with an intense fitness regimen.

Fortunately, he developed a new social network of other like-minded and intrepid teammates who shared his same hunger to change the world—while living and working off-world. In their downtime, they talked about how they saw their collaboration as something like what a crew of seafaring explorers might have felt centuries ago, surveying remote stretches of the North Pole or charting the Great Barrier Reef.

Choosing to look at their circumstances in this way steeled their resolve when the work got too hard or the isolation too intense. It heartened them, too, making them feel as if they were a part of something important, something bigger than themselves.

Daily, Angel served on the team tasked with building solar mirrors in space. Bound for Mars, they were intended to capture tremendous amounts of energy to be pointed at the planet's polar ice caps. "Melting these will unlock precious water that's been locked up in ice for eons," his supervisor told Angel and his team. "Just imagine. The work you're doing today will once again restore Mars's great rivers—its seas. Because of you, the first Martian colony will be able to tap into its own water supply, making the planet more hospitable, if not *lush* with vegetation—generations from now."

The key idea is *thinking generationally*, something not enough of today's Earth's denizens tend to do. Our economy is built upon short-term gains and instantaneous gratification, affording little patience for such contemplation.

Yet if we hope to inspire the brightest minds and greatest talents to seize the opportunity space affords, this is the required mentality. We are at the doorstep of unlocking the next generation of exploration. What's needed now is inspiring those pioneers with a grand vision, stirring them to action.

Why? For all that humanity has discovered and learned about space, fewer than 600 people have reached Earth's orbit. Sierra Space and other commercial trailblazers wish to change that. We envision a future where humanity lives and works in space, on moons, and on distant planets—a tomorrow where we become a Type I civilization and beyond, forever pushing humankind against the limits of possibility.

It all begins with real people like our fictional Angel, those who have the right stuff, those who know there is more out there in the final frontier. Does that sound like you? If so, space could use you. *We* could use you.

CARTOONIST BILL WATTERSON IS best known for creating *Calvin and Hobbes*, a comic about an inventive boy (Calvin) and his imaginary toy tiger pal (Hobbes). It became a beloved favorite for kids and adults alike ever since Universal Press Syndicate bought the strip in 1985, affording then-27-year-old Watterson a national audience.

Fans love Calvin's funny antics and precocious observations of life. They also enjoy the deep friendship between him and Hobbes—not to mention Calvin's alter ego: Spaceman Spiff, a zany astronaut who journeys the cosmos, crash-landing on hostile worlds and fighting off aliens.

Yet after a decade of producing the comic, Watterson retired it on December 31, 1995, to the dismay of millions. The final book of comics Watterson produced is called *It's a Magical World*. The very last panel on the last page of the book ends on an optimistic note. We would like to conclude our own book with its message, as it captures the hopefulness so needed to carry our species into the great unknown, not just with determination, but with unbridled *joy*.

On the final page, Calvin and Hobbes wear scarves and mittens. Carrying their toboggan, they marvel at how an endless blanket of snow has transformed everyday life into something glorious—a playground of childhood delight. Their simple, earnest exchange below mirrors our thoughts on the future to come: the mentality we wish to cultivate among the people of Earth.

Hobbes: "Everything familiar has disappeared! The world looks brand new!"

Calvin: "A day full of possibilities! It's a magical world, Hobbes, ol' buddy . . . Let's go exploring!"[15]

Appendix

COOKING IN SPACE RECIPES

Prepared by Chef Julian Martinez

SPHERIFIED NEGRONI WITH ORANGE-ROSEMARY SUGAR

NEGRONI SPHERES

130 grams sweet vermouth

130 grams gin

130 grams Campari

Zest of 1 orange

4 sprigs rosemary

7.5 grams calcium lactate

1 gram xanthan gum

0.2 gram ascorbic acid

5 grams sodium alginate

1000 grams water

Orange-rosemary sugar (see recipe on the following page)

On Earth:

1. Combine the sweet vermouth, gin, Campari, orange zest, and the whole sprigs of rosemary in a container or lidded jar. Let sit at least 8 hours, or overnight.

2. Strain the alcohol mixture, discarding the solids. Blend alcohol mixture with the calcium lactate, xanthan gum, and ascorbic acid. Refrigerate until ready to use.

3. Blend the sodium alginate and water until completely dissolved. Refrigerate the solution for 1 hour.

4. Fill a tablespoon with the alcohol mixture, then tip it slowly into the alginate bath, releasing the alcohol into the alginate solution while retaining a spherical shape. Let sit for 7 minutes, allowing a skin to form around the alcohol.

5. Remove the negroni sphere from the bath with a slotted spoon. Dip the sphere into a second bath of plain water to rinse.

6. Place negroni spheres in a lidded ice cube tray or silicone mold. Sprinkle each sphere with orange-rosemary sugar (see recipe below), then cover the tray or the silicone mold with a clasping lid or tightly with plastic wrap and refrigerate.

ORANGE-ROSEMARY SUGAR

Zest of 4 oranges

100 grams sugar

2 sprigs rosemary, minced

50 grams water

On Earth:

1. Combine the orange zest, sugar, minced rosemary, and water in a small saucepan. Heat over medium-low heat, stirring, until the sugar dissolves.

2. Continue to cook the sugar mixture until the water has evaporated completely. Transfer sugar to a plate to let cool.

To serve on orbit (serves 6):

6 negroni spheres

To serve negroni spheres, remove the tray or silicone mold from the refrigerator. Present to guests and serve immediately.

THE CHIP AS THE DIP

AGED CHEDDAR PUFF

50 grams highest-quality aged cheddar cheese

100 grams tapioca flour

5 grams kosher salt

75 grams water

Rice bran oil, for frying

On Earth:

1. Finely grate the cheese with a microplane. Transfer cheese to a food processor. Add the tapioca flour and salt. Turn on the processor and slowly add the water, in a thin stream. A dough-like mixture should form.

2. Remove the dough from the processor and gently spread out between two sheets of plastic wrap. Roll the dough out with a rolling pin to about 3 mm thickness.

3. With the dough still between the two sheets of plastic, transfer to a steamer basket or steam oven. Steam on high heat for 12–13 minutes on the first side, flip, then steam an additional 12–13 minutes on the second side. Remove from steamer and let cool completely. Remove the plastic wrap.

4. Transfer the steamed dough sheets to a dehydrator and dehydrate at 135°F for 3 hours.

5. Heat oil to 425°F. Fry the cheese sheets for 1½–2 minutes, or until completely puffed. Let drain on a paper towel–lined tray.

SAVORY CORN PUDDING

Kernels from 5 ears yellow corn

Salt

Juice from 1 lime

Cornstarch, as needed

On Earth:

1. Juice the corn using a masticating juicing machine. Pass through a fine mesh sieve.

2. Pour corn juice into a wide pan. Heat over medium heat, whisking constantly. Don't let the juice burn. As the juice heats up, it will thicken due to the natural starches present in the corn. Once it is creamy like mayo, remove from heat and add the lime juice. Season to taste with salt. If the corn juice just doesn't want to thicken, add a couple of teaspoons of cornstarch and whisk in until it does thicken. Let cool, then transfer to a squeeze bottle for storage.

TOMATO POWDER

5 tomatoes, sliced thinly

On Earth:

1. Spread the tomatoes on dehydrator sheets. Place in a dehydrator, set to 120°F. Let dehydrate overnight, or until the tomatoes are completely brittle and dried out. Remove from the dehydrator and let cool.

2. Blend dried tomatoes in a blender until pulverized. Pass through a fine mesh sieve to make sure no large bits remain.

🚀 To serve on orbit (serves 6):

12 aged cheddar puffs

1 small squeeze bottle corn pudding, warmed

1 small shaker tomato powder

Making sure to keep the puffs contained as you assemble them, squeeze a generous dollop of corn pudding onto each cheese puff. Top with tomato powder and present to diner immediately.

SPLIT PEA S'MORES

SPLIT PEA PUREE

15 grams grapeseed oil

150 grams carrots, thinly sliced

150 grams leeks, coarsely chopped

150 grams onions, chopped

5 grams salt

25 grams unsalted butter

1000 grams chicken stock

1 smoked ham hock

1 California bay leaf

325 grams green split peas, rinsed

150 grams fresh green peas, blanched

¼ bunch Italian parsley

10 grams white wine vinegar

Freshly ground black pepper

On Earth:

1. Heat the grapeseed oil in a stockpot over medium heat. Add the carrots, leeks, onions, and salt. Add the butter and let melt. Reduce the heat to low and cover the vegetables with a parchment paper lid with a hole in the center. Cook for about 30 minutes, or until the vegetables are very tender but have no color.

2. Add the chicken stock, ham hock, and bay leaf to the pot. Bring to a simmer, then cook, covered, on medium-low for 1 hour. Strain the stock into a bowl. Discard the used vegetables and bay leaf and

return the stock to the pot, along with the ham hock. Add the split
peas and bring the stock to a simmer. Let simmer for 1 hour, or
until the peas are completely softened.

3. Remove the soup from the heat. Remove and discard the ham
   hock. Add the blanched peas and the parsley. Puree well in a high-
   speed blender until very smooth, then pass the puree through a
   fine-mesh strainer. Season with the white wine vinegar, additional
   salt, and fresh pepper to taste.

HOT SPLIT PEA MARSHMALLOW

500 grams Split Pea Puree, from above

7 grams methylcellulose F50

15 grams 160 Bloom strength gelatin powder

On Earth:

1. Combine the methylcellulose with 400 grams split pea puree in a
   saucepan. Reserve the remaining 100 grams of puree for later use.
   Bring pea puree mixture to a boil and then simmer for 2 minutes.
   Remove from heat and let cool completely. Transfer to refrigerator
   and let sit overnight.

2. Combine the gelatin and the remaining split pea puree in a small
   saucepan. Bring to a simmer to dissolve. Combine the gelatin-
   infused pea puree with the methylcellulose-infused pea puree.
   Transfer mixture to the bowl of a stand mixer fitted with a whisk
   attachment. Whisk on high for 8–10 minutes, or until light and airy.

3. Transfer the mixture to a square-shaped oven-proof silicone mold.
   Fill each compartment of the mold with the puree, leveling out the
   top with an offset spatula. Cover the mold with plastic wrap and
   refrigerate for 4–6 hours, or until set.

4. Bake the marshmallows at 250°F for 3–4 minutes. Remove from
   oven and let cool in the mold. Tightly tape magnets to the bottom
   of the mold, which will be needed in space for securing the mold

in place. Cover the mold with fresh plastic wrap, then store in a secure space for the trip to space.

PUMPERNICKEL CRACKER

1 loaf pumpernickel bread

100 grams unsalted butter, melted

Maldon salt

On Earth:

1. Place the loaf of pumpernickel in the freezer for about 2 hours, or until mostly frozen solid.

2. Using a deli slicer, cut the bread into thin slices, about ½ cm in thickness. Cut each slice into a neat square, about 4" by 4".

3. Lay the bread slices on a sheet tray. Spread with melted butter, then season with Maldon salt. Bake at 300°F for about 15 minutes, or until crispy but not burnt. Remove from oven and let cool.

To serve at on orbit (serves 6):

6 split pea marshmallows

6 thin slices pork lardo

12 pumpernickel crackers

1. Working quickly so as to not let the lardo float away, top one of the marshmallows (still in the mold) with a slice of lardo. Transfer to radiation warming oven and warm for about 30 seconds on medium heat.

2. Once out of the oven, transfer the mold of marshmallows to a magnetized surface.

3. Working with two crackers at a time, quickly assemble each s'more by sandwiching one warm marshmallow between two pumpernickel crackers. Press together to ensure the marshmallow

sticks to the crackers without floating apart. Serve each s'more immediately.

SALAD IN AN ENVELOPE

ORANGE BLOSSOM ENVELOPE

4 sheets silver gelatin

675 grams orange blossom water

125 grams filtered water

8 grams pectin

12 grams agar

1 gram kosher salt

3 grams citric acid

120 grams sugar

20 grams sorbitol

On Earth:

1. Place the gelatin sheets in ice water for 4–5 minutes, or until softened. Squeeze out as much water as possible, then set aside.

2. In a blender, combine the orange blossom water, water, pectin, agar, salt, citric acid, sugar, and sorbitol. Blend well. Transfer to a saucepan and bring to a boil over medium heat. Add the hydrated gelatin to the pan and stir until dissolved. Pour the liquid into a medium bowl, cover with plastic wrap, and refrigerate until set, 2–3 hours.

3. Transfer the set gel to a blender. Blend on high, pushing down the sides to ensure everything gets blended well. Pass through a fine mesh sieve.

4. Cut 8 pieces of acetate plastic 3" by 12". Carefully spread the pureed gel onto each strip. Transfer the strips to a dehydrator and dry at 90°F for 6–8 hours, or until turned into a leather.

5. Remove the leather from the plastic and cut the sheets into 3" by 4" rectangles, making sure the edges are squared.

6. Fold the long end of each rectangle over to meet the opposite long end. Using a heat sealer, seal that end together. Then seal one of the short ends together. Reserve envelopes for filling.

BEET BRUNOISE

2 large orange beets

500 grams orange juice

500 grams water

10 grams salt

On Earth:

1. Combine the orange juice, water, and salt in a saucepan. Stir to combine, then add the beets. Cover with a lid and cook on medium-low heat for about 45 minutes, or until beet is fork tender. Do not overcook the beets, as you want them to have a bit of texture. Remove from liquid.

2. Let beets cool slightly, then peel skin away. Place beets in refrigerator to chill completely.

3. Cut beets into a fine brunoise.

GRAPEFRUIT VESICLES

1 grapefruit

400 grams olive oil

On Earth:

1. Cut the grapefruit into segments. Submerge in olive oil. Stirring occasionally, let the segments sit in the oil for 15–20 minutes, or until the segments break down. Strain through a fine mesh sieve, reserving the oil for the olive oil powder. Rinse the grapefruit vesicles under cold water. Store in the refrigerator.

BRIOCHE CROUTONS

30 grams olive oil

2 cloves garlic, smashed

150 grams brioche bread, crusts removed, diced into ¼-inch cubes

5 sprigs thyme

Salt

On Earth:

1. Heat the olive oil in a sauté pan over medium heat. Add the garlic cloves. When the garlic begins to sizzle, add the cubed brioche and cook, tossing occasionally, until the bread begins to turn golden, about 5 minutes. Make sure it doesn't burn. Toward the end of the cooking, add the thyme and continue to toss for 30 seconds or so.

2. Once the bread is golden brown, immediately remove from heat, transfer to a baking sheet lined with paper towels, and season with salt to taste. Remove the garlic and thyme. Once cooled, store in an airtight container.

WHITE BALSAMIC MINI SPHERES

200 grams white balsamic vinegar

4 grams agar

1 quart olive oil, very well chilled

On Earth:

1. Bring the white balsamic vinegar and agar to a boil. Boil for 1 minute to dissolve the agar. Make an ice bath. Then place the oil, set in a plastic quart container, in the ice bath.

2. Extract the vinegar mixture with a squeeze bottle, then slowly drip into the cold oil, drip by drip. Swirl the container of oil as you drip the vinegar mixture into it, creating a small vortex effect. Continue until the vinegar is done. Strain the oil through a fine mesh sieve, catching the "caviar" balls. Reserve the oil for another use.

POWDERED OLIVE OIL

5 grams maltodextrin

50 grams reserved olive oil (from Grapefruit Vesicles)

Kosher salt

On Earth:

1. Place the maltodextrin in a food processor. Turn to low speed.

2. Gradually pour in the olive oil, keeping a close eye on it to make sure it remains a powder rather than turning to a solid clump.

3. Once the oil is incorporated and the mixture remains fluffy, remove from the food processor. Season with salt to taste.

SALAD ENVELOPES

6 orange blossom envelopes

1 pint micro-lettuce mix

Beet brunoise

White balsamic spheres

Brioche croutons

Powdered olive oil

Grapefruit vesicles

Aged goat gouda, shaved into ribbons with vegetable peeler

50 grams pistachios, chopped

On Earth:

1. Carefully spread one orange blossom envelope open. Carefully place some lettuce mix in the bottom. Then add about one tablespoon each of the beet brunoise and white balsamic spheres. Add a bit more of the micro-lettuces. Then add about one tablespoon each of the brioche croutons and powdered olive oil. Add a bit more micro-lettuces. Finish by adding about a teaspoon of grapefruit vesicles, a few shavings of goat gouda, and a pinch of chopped pistachios.

2. Using a heat sealer, seal the open end of the envelope. Repeat process with remaining envelopes. Store packed envelopes in a secure container.

🚀 To serve on orbit (serves 6):

6 salad envelopes

1. Carefully remove each packed envelope from refrigeration. Present each to diners immediately.

KOREAN BBQ ON A CRACKER

SMOKED NORI LAVASH CRACKER

100 grams buckwheat flour

160 grams bread flour

20 grams wheat germ

7 grams kosher salt

10 grams smoked nori, coarsely ground

160 grams water

40 grams sourdough starter

1. Combine the buckwheat flour, bread flour, wheat germ, salt, and smoked nori in a bowl. Mix well.

2. Combine the water with the sourdough starter. Mix to disperse. Add the water-sourdough mixture to the flour mixture. Mix well. Form into a ball and let ferment in the refrigerator overnight.

3. The next day, preheat oven to 425°F. Divide the dough into 50-gram portions. Shape into small balls and let rest 15 minutes. Coat a work surface with flour and place one dough piece on the surface. Sprinkle flour on top. Using a rolling pin, roll the dough as thinly as possible. Flip, coating with more flour, and continue to roll until the dough is evenly thin.

4. Cut the dough into 6" by 2" rectangles with even edges. Transfer the rectangles to a parchment-lined sheet tray and bake until golden-brown, about 8–10 minutes.

5. At this point, the crackers will be mostly cooked, but won't be crispy. To make sure they are extra crisp, transfer to a dehydrator set at 180°F for 30–60 minutes. Let cool.

GOCHUJANG MAYO

120 grams high-quality or homemade mayonnaise

20 grams gochujang paste

1. Whisk together the mayonnaise with the gochujang. Store sauce in a squeeze bottle or culinary syringe.

KIMCHI PUREE

400 grams kimchi

15 grams sesame oil

1. Combine the kimchi and sesame oil in a food processor. Blend very well, adding more sesame oil if needed to thin out the puree. Store kimchi puree in a wide-mouthed squeeze bottle.

SHISO PESTO

100 grams yellow miso

25 grams raw pistachios

30 grams shallots, chopped

15 grams garlic, chopped

75 grams olive oil

25 grams cold water

50 grams yuzu juice

25 grams lemon juice

3 grams salt

1 gram ground white pepper

1 bunch shiso

1. Combine all of the ingredients except the shiso in a blender and blend on high until smooth. Add the shiso and quickly blend on high speed to pulverize the herbs, without heating the puree. Season to taste with additional salt and lemon juice. Store in a squeeze bottle.

GELLED GINGER

200 grams ginger juice
3 grams agar

1. In a saucepan, combine the ginger juice and agar. Bring to a boil, then whisk for about 20 seconds. Remove from heat and transfer to a bowl. Cover with plastic wrap and refrigerate until set, about 1 hour.

2. Place the brittle set gel into a blender. Blend on high until it breaks down into a fluid gel. You may have to push the gel down the sides with a ladle to get it moving. You can also add more unset ginger juice if you need to help it get going. Once fully blended, pass through a fine-mesh strainer and store in a squeeze bottle in the refrigerator for up to 5 days.

To serve on orbit (serves 6):

6 Smoked Nori Lavash Crackers
6 small squeeze bottles Gochujang Mayo
6 small squeeze bottles Shiso Pesto
6 small squeeze bottles Kimchi Puree
6 small squeeze bottles Gelled Ginger
100 grams wagyu beef sirloin, sliced paper-thin,
   stored in vacuum-sealed bag

1. Distribute one smoked nori cracker to each diner. Instruct diners to make sure to hold the cracker while walking them through self-assembly of the dish.

2. Distribute a small squeeze bottle of gochujang mayo to each diner. Instruct diners to pipe 4 evenly spaced lines gochujang mayo onto the surface of the cracker, going straight across the width of the cracker.

3. Collect the squeeze bottles of gochujang mayo and exchange for a bottle of shiso pesto to each diner. Instruct diners to pipe one line of the pesto right next to the lines of mayo. Repeat this with the kimchi puree and gelled ginger. The diners should end up with sixteen lines across each cracker, with 4 alternating lines of each flavorful component.

4. Using tweezers, distribute 3 pieces of thinly sliced beef sirloin, gently splaying atop each diner's cracker. Instruct diners to eat cracker immediately, making sure to contain all crumbs of the crackers to avoid rogue cracker bits floating around for the remainder of the meal.

SAVORY FROZEN MOCHI

SHABAZI SPICE BLEND

15 grams jalapeno powder

5 grams garlic powder

40 grams cilantro powder

30 grams ground parsley

5 grams ground cardamom

10 grams ground cumin

10 grams lemon peel powder

5 grams Aleppo chili powder

3 grams citric acid

2 grams ground black pepper

1. Combine all ingredients in a dry blender. Blend very well. Pass through a fine mesh sieve.

SHABAZI-SPICED MOCHI

300 grams Mochiko sweet rice flour

150 grams sugar

5 grams Shabazi spice blend

350 grams water

400 grams cornstarch, for dusting

On Earth:

1. In a microwave-safe bowl, mix together the sweet rice flour, sugar, and Shabazi spice blend. Add the water and mix everything until homogeneous.

2. Cover the bowl with plastic wrap and microwave for 2 minutes. Stir contents of the bowl and microwave an additional 30–60 seconds, or until the dough is slightly translucent.

3. Place a sheet of parchment paper on the countertop. Dust the parchment with cornstarch. Transfer the rice dough to the parchment and sprinkle with more cornstarch.

4. Using a dough scraper, divide the dough into 6 equal parts. Flatten each portion into a disc, about 1/8 inch in thickness. Sprinkle a rectangular-shaped silicone mold (4" x 2") with cornstarch. Lay the dough discs in the mold. Cover with plastic wrap until ready to assemble the mochi cakes.

FOIE GRAS MOUSSE

200 grams chicken livers, cleaned and soaked in milk overnight

100 grams foie gras, vein removed, cut into 1-inch pieces

50 grams thinly sliced shallots

Salt, to taste

50 grams Madeira

100 grams unsalted butter, softened at room temperature

150 grams chicken stock

On Earth:

1. Heat a pan over high heat. Add the foie gras and sear, being careful not to let too much of the fat cook off. Transfer to wire rack set over a sheet pan.

2. Decrease the heat to medium and add the shallots, seasoning with a pinch of salt. Sweat until softened. Deglaze with the Madeira. Add the chicken livers and cook to medium doneness, about 5 minutes. Let cool.

3. To a blender, add the foie gras and the liver mixture. Blend, alternating between adding the butter and chicken stock. Season mixture to taste with salt.

4. When emulsified, pass the mixture through a fine mesh sieve into a bowl set in an ice bath. Chill. Transfer to a piping bag and store in the refrigerator until ready to assemble.

CHERRY GASTRIQUE

450 grams sweet cherries, halved and pitted

50 grams sugar

90 grams sherry vinegar, divided

Kosher salt

Freshly ground black pepper

On Earth:

1. Combine the cherries with sugar and 60 grams of sherry vinegar. Cook over medium heat, stirring frequently, until the sugar is dissolved and the cherries have cooked down to a jam-like consistency. Season with salt and pepper.

2. Add remaining 30 grams of vinegar. Simmer sauce for an additional 10 minutes, if necessary, to reduce to a lightly syrup-like consistency. Let cool. Store in a squeeze bottle.

CHILI HAZELNUTS

200 grams blanched hazelnuts

20 grams honey

15 grams olive oil

3 grams Aleppo chili powder

On Earth:

1. Preheat the oven to 300°F. Combine the hazelnuts, honey, olive oil, and chili powder in a bowl. Mix well. Transfer nuts to a sheet tray and bake until nuts are lightly toasted and fragrant.

2. Let nuts cool. Then chop roughly.

ASSEMBLY

6 pieces Shabazi-Spiced Mochi

Chili Hazelnuts

Maldon salt

Foie Gras Mousse

Cherry Gastrique

1 loaf brioche bread, cut into ½-inch pieces

On Earth:

1. Tightly press one mochi disc into a compartment of the rectangular silicone mold, letting the excess mochi hang off to the sides. Sprinkle the surface of the mochi evenly with chili hazelnuts, then sprinkle a pinch of Maldon salt on top. Then carefully pipe an even line of the foie gras mousse across the length of the cake. Squeeze a layer of cherry gastrique atop the mousse.

2. Cut the brioche slices into rectangles of the same size as the mold. Place a brioche slice atop the cherry gastrique in the mold. Fold the excess mochi toward the center of the cake. Seal the corners together, meeting at the center.

3. Remove the mochi cake from the mold and turn over. Repeat with the remaining pieces of mochi and the remaining ingredients. Store mochi cakes in the freezer.

🚀 To serve on orbit (serves 6):

6 savory mochi cakes

1. Remove the mochi cakes from zero-gravity freezer. Let sit for 3–5 minutes in an enclosed container, to take off a bit of the chill. Present cakes to diners, allowing them to retrieve the cakes from the air.

DULCE DE LECHE SNOWBALL

CAJETA CARAMEL FUDGE

115 grams unsalted butter

400 grams light brown sugar

400 grams sweetened condensed milk

10 grams vanilla extract

On Earth:

1. Line a quarter sheet tray with parchment paper and spray with cooking spray. Set aside.

2. Combine the butter, brown sugar, sweetened condensed milk, and vanilla extract in a saucepan. Bring to a boil over low heat. Simmer for 15 minutes, stirring frequently.

3. After the initial cooking phase, test the fudge by placing a teaspoon of the mixture in a bowl with cold water. Drain the water and mold the fudge into a ball. If the ball is able to retain its shape, the fudge is ready. If not, continue to simmer over low heat for 2–3 more minutes, then test the fudge again.

4. Once the fudge passes the test, transfer the contents of the pan to the bowl of a stand mixer. Mix on medium-low for 5 minutes. The fudge will begin to set. Immediately transfer to the prepared pan and spread it out with an offset spatula. Refrigerate for at least 2 hours.

5. Cut the fudge into ½-inch squares, then roll them into ball shapes. They should be the size of a standard marble. Store in the refrigerator until ready to assemble cheesecake balloons.

CHEESECAKE FILLING

500 grams full-fat cream cheese

115 grams goat cheese

8 grams lemon juice

5 grams vanilla extract

1 gram kosher salt

2 grams orange blossom water

200 grams sugar

2 large eggs

250 grams heavy cream

On Earth:

1. Combine the cream cheese, goat cheese, lemon juice, vanilla extract, salt, and orange blossom water in the bowl of a stand mixer fitted with a paddle attachment. Mix on low speed until the mixture is roughly combined. Increase the speed to medium and mix until no lumps of cheese remain, about 5 minutes.

2. Reduce the speed of the mixer to medium-low and add the sugar. Mix until just combined. Remove the bowl from the stand mixer and set a fine mesh sieve over the top. Crack the eggs into the sieve and whisk until they pass through. Discard any bits of egg that will not pass through the strainer easily.

3. Return the mixture to the stand mixer and resume mixing until combined.

4. Heat the cream to a boil in a saucepan. While the mixer is still running, slowly add the cream in a steady stream. Mix until just combined.

5. Remove the cheese mixture from the mixing bowl into the canister of a whipping siphon. Let chill completely. Charge with two NO_2 charges and reserve until assembly of cheese snowballs.

SNOWBALL ASSEMBLY

1 charged canister Cheesecake Filling
6 Cajeta Caramel fudge balls
Smoked Maldon salt
6 water balloons

On Earth:

1. Invert the whipped cream canister and shake vigorously for 30 seconds. Inflate one water balloon halfway with the cheesecake filling. Then push one cajeta caramel ball into the balloon, doing your best to place it as close to the center of the filling as possible. Sprinkle in a small pinch of smoked Maldon salt inside the balloon. Fill the balloon with more cheesecake filling. The ball should be about the size of a lacrosse ball. Release any excess air from the balloon and tie tightly. Place the balloon gently aside. Repeat with remaining balloons. Balloons can be stored in a Styrofoam wine case secured in place in the refrigerator on the space shuttle.

DULCE DE LECHE

1 can sweetened condensed milk
Kosher salt, to taste

On Earth:

1. Place an unopened can of sweetened condensed milk in a pot filled with water. Bring to a boil, cover the pot, and cook the can at a full boil for 2 hours.

2. Let cool completely, then open can and stir in a pinch of salt. Divide caramelized milk between 6 miniature squeeze bottles. Store in the refrigerator.

To serve on orbit (serves 6):

6 cheesecake-filled balloons

6 miniature squeeze bottles of Dulce de Leche

1. Set up a liquid nitrogen bath. Working one at a time, place balloons in the liquid nitrogen using a clamp to hold the balloon. Spin balloon around for about 30 seconds, or until the balloon feels solid.

2. Remove the balloon from the nitrogen and cut off the tied balloon. Let sit for 1 minute, without letting it float away. Use small tweezers to peel away the balloon. Store the finished frozen snowball in the secured Styrofoam wine case in the refrigerator and repeat process with the remaining balloons.

3. Present the snowballs to diners along with one squeeze bottle of dulce de leche. Instruct diners to hold on to the snowball in one hand and the squeeze bottle in the other, squeezing the dulce de leche directly onto the snowball with each bite.

Notes

CHAPTER 1

1. Carl Sagan, *Pale Blue Dot: A Vision of the Human Future in Space* (New York: Ballantine Books, 1997).

CHAPTER 2

1. Nathaniel Lee and Jessica Orwig, "What It Takes to Become an Astronaut," *Business Insider*, October 24, 2018, video, 0:58, https://www.businessinsider.com /nasa-astronaut-explains-space-training-2018-10.

2. Buckminster Fuller, "Accelerating Acceleration," Buckminster Fuller Institute, *Trimtab Blog*, accessed May 2, 2023, https://www.bfi.org/resource/accelerating -acceleration/.

3. Zaryn Dentzel, *Ch@nge: 19 Key Essays on How the Internet Is Changing our Lives* (Nashville: Turner, 2014).

4. "Running the Race to Cure Cancer From Space," October 1, 2014, https: //www.nasa.gov/mission_pages/station/research/news/cancer_research_in_space.

5. "Four Decades of Cancer Research in Space: How We've Benefited," April 12, 2013, https://www.medicaldaily.com/four-decades-cancer-research-space-how-weve -benefited-245002.

6. Rich Miller, "Server Farms Get Super-Sized for Cloud Growth," *Data Center Knowledge*, April 21, 2014, https://www.datacenterknowledge.com/ archives/2014/04/21/server-farms-get-super-sized-cloud-growth.

7. Nikitha Sattiraju, "The Secret Cost of Google's Data Centers: Billions of Gallons of Water to Cool Servers," *Time*, April 2, 2020, https://time.com/5814276/ google-data-centers-water/.

8. Maria Korolov, "Space Is the Final Frontier for Data Centers," *Data Center Knowledge*, January 18, 2022, https://www.datacenterknowledge.com/hardware/ space-final-frontier-data-centers.

9. James Wall, "Ice Ice Baby—Why Quantum Computers Have to Be Cold," *The Quantum Authority*, December 23, 2017, https://medium.com/thequantum-authority/ice-ice-baby-why-quantum-computers-have-to-be-cold-3a7f777d9728.

10. "What Is Quantum Computing?" IBM, accessed May 2, 2023, https://www.ibm.com/topics/quantum-computing.

11. European Space Agency, "Quantum Technologies in Space: Policy White Paper," Cosmos ESA, August 2019, https://www.cosmos.esa.int/documents/1866264/3219248/BassiA_QT_In_Space_-_White_Paper.pdf/6f50e4bc-9fac-8f72-0ec0-f8e030adc499?t=1565184619333.

12. Mike Cruise, "Factories in Space: How Extra-Terrestrial Industry Could Keep Humans Alive," *The Conversation*, August 18, 2016, https://theconversation.com/factories-in-space-how-extra-terrestrial-industry-could-keep-humans-alive-63548.

13. Michael Sheetz, "Super Fast Travel Using Outer Space Could Be $20 Million Market, Disrupting Airlines, UBS Predicts," CNBC, March 18, 2019, https://www.cnbc.com/2019/03/18/ubs-space-travel-and-space-tourism-a-23-billion-business-in-a-decade.html#:~:text=In%20a%20decade%2C%20high%20speed,to%20%24805%20billion%20by%202030.

14. "How Is the Entertainment Industry Helping to Spark People's Interest in Space Travel?," *Space Coast Daily*, March 10, 2021, https://spacecoastdaily.com/2021/03/how-is-the-entertainment-industry-helping-to-spark-peoples-interest-in-space-travel/.

15. K. J. Yossman, "New Film Studio Will Be Built in Space by 2024," *Variety*, January 19, 2022, https://variety.com/2022/film/news/film-studio-space-1235157521/.

16. Lexie Cartwright, "Tom Cruise Set to Become First Actor to Shoot Movie in Outer Space," *New York Post*, October 9, 2022, https://nypost.com/2022/10/09/tom-cruise-set-to-become-first-actor-to-shoot-movie-in-outer-space/.

CHAPTER 3

1. Plutarch, *Life of Theseus* 23.1.

2. Poul Anderson, *The Boat of a Million Years* (New York: Tor Books, 1989).

CHAPTER 4

1. NASA Content Administrator, ed., "What Is Microgravity?," NASA, last modified August 17, 2017, https://www.nasa.gov/centers/glenn/shuttlestation/station/microgex.html.

2. Tommy Brooksbank et al., "International Space Station Caught in Crosshairs of Geopolitical Tensions," ABC News, March 9, 2022, https://abcnews.go.com/International/us-russian-international-space-station-partnership-jeopardy-geopolitical/story?id=83343874.

3. "Astronaut-Turned-Artist Nicole Stott Shares View from Space in Paintings," Collect Space, July 8, 2016, http://www.collectspace.com/news/news-070816a-astronaut-artist-nicole-stott.html.

4. Eldora Valentine, "Race from Space Coincides with Race on Earth," NASA, March 17, 2007, https://www.nasa.gov/mission_pages/station/expeditions/expedition14/exp14_boston_marathon.html.

5. "Your Body in Space: Use It or Lose It," NASA, August 8, 2004, https://www.nasa.gov/audience/forstudents/5-8/features/F_Your_Body_in_Space.html.

6. Mike Massimino, "'I'm an Astronaut. This is What It's Like to Walk in Space,'" *Newsweek*, May 5, 2020, https://www.newsweek.com/im-astronaut-this-what-space-travel-like-1501881.

7. Massimino, "'I'm an Astronaut.'"

CHAPTER 5

1. Brant Cox, "Tokyo Delve's Sushi Bar," Reviews, *The Infatuation*, February 9, 2017, https://www.theinfatuation.com/los-angeles/reviews/tokyo-delves-sushi-bar.

2. Ishay Govender-Ympa, "Dining in the Dark: World's Blind Restaurant Experiences," *Fine Dining Lovers*, June 24, 2014, https://www.finedininglovers.com/article/dining-dark-worlds-blind-restaurant-experiences.

3. Rachel Hommel, "Chef's Corner: Julian Martinez," *Santa Barbara Independent*, April 5, 2016, https://www.independent.com/2016/04/05/chefs-corner-julian-martinez/.

4. Patrice Novotny, "A Japanese Maitre D' Is Officially the Best Waiter in the World," *Business Insider*, November 12, 2012, https://www.businessinsider.com/a-japanese-maitre-d-is-officially-the-best-waiter-in-the-world-2012-11.

5. Kate Boland, "Chinese Food Habits That Westerners Will Never Understand," World Atlas, March 15, 2020, https://www.worldatlas.com/articles/chinese-food-habits-that-westerners-will-never-understand.html.

6. Amy Bingham, "What Are Tapas? A Guide to Spain's Small Plates," *Spanish Sabores* (blog), March 9, 2021, https://spanishsabores.com/spain-dining-guide-what-are-tapas/.

7. Danilo Alfaro, "How Does an Air Fryer Work?," *The Spruce Eats*, January 25, 2023, https://www.thespruceeats.com/how-does-an-air-fryer-work-4693673.

8. Angie Bates, "What Is Spherification?," Delighted Cooking, March 21, 2023, https://www.delightedcooking.com/what-is-spherification.htm.

9. "The Use of Liquid Nitrogen in Food and Drinks," Food Alert, February 26, 2015, https://www.foodalert.com/use-liquid-nitrogen-food-and-drinks.

10. Allen Hemberger, "Granola in a Rosewater Envelope," The Alinea Project, https://www.allenhemberger.com/alinea/2009/05/granola-in-a-rosewater-envelope/.

11. Vanessa Greaves, "What Is Mochi and How Do You Make It at Home?," All Recipes, February 16, 2021, https://www.allrecipes.com/article/all-about-mochi/.

CHAPTER 7

1. Ernest Hemingway, *A Moveable Feast* (New York: Scribner, 2010).

2. Anne Rice, *Interview with the Vampire* (New York: Ballantine Books, 1991).

CHAPTER 10

1. Christine Comaford, "New Research on the Real Cause of The Great Resignation," *Forbes*, January 14, 2022, https://www.forbes.com/sites/christinecomaford/2022/01/14/new-research-on-the-real-cause-of-the-great-resignation/?sh=5172a98b3b72.

2. Laura Conover, "How to Become an Employer of Choice in 2021," LinkedIn, April 27, 2021, https://www.linkedin.com/pulse/how-become-employer-choice-2021-laura-conover/.

3. Gaby Galvin, "Nearly 1 in 5 Health Care Workers Have Quit Their Jobs During the Pandemic," *Morning Consult*, October 4, 2021, https://morningconsult.com/2021/10/04/health-care-workers-series-part-2-workforce/.

4. Betsy Reed, "'Exhausted and Underpaid': Teachers across the US Are Leaving Their Jobs in Numbers," *The Guardian*, October 2021, https://www.theguardian.com/world/2021/oct/04/teachers-quitting-jobs-covid-record-numbers.

5. Dianne Hermann, "Police Officers Are Retiring and Quitting at Record Levels," *Independent Sentinel*, April 23, 2021, https://www.independentsentinel.com/police-officers-are-retiring-and-quitting-at-record-levels/.

6. Hillary Hoffower, "65% of Gen Zers Plan to Join the Great Resignation This Year, Survey Finds," *Business Insider*, February 4, 2022, https://www.businessinsider.com/gen-z-quitting-joining-great-resignation-reshuffle-job-with-purpose-2022-2.

7. Ashutosh Gupta, "What Is a Metaverse?," Gartner, January 28, 2022, https://www.gartner.co.uk/en/articles/what-is-a-metaverse.

8. Allison Goldberg, "How Media Changes the Way We Communicate," *Forbes*, November 7, 2017, https://www.forbes.com/sites/quora/2017/11/07/how-media-changes-the-way-we-communicate/?sh=4f1777d72f5b.

9. Harry L. Katz, *Mark Twain's America: A Celebration in Words and Images* (New York: Little, Brown and Company, 2014).

10. Ashley Grice, "The Great Resignation Is upon Us and Purpose Can Help You Resist Its Siren Call," *Fortune*, September 30, 2021, https://fortune.com/2021/09/30/great-resignation-leading-with-purpose-bcg-brighthouse/.

11. Jolene Creighton, "The Kardashev Scale - Type I, II, III, IV & V Civilization," *Futurism*, August 21, 2019, https://futurism.com/the-kardashev-scale-type-i-ii-iii-iv-v-civilization.

12. "The Levels of Civilization in the Universe (Kardashev Scale)," Science Info, https://scienceinfo.net/the-levels-of-civilization-in-the-universe-kardashev-scale.html.

13. Douglas Adams, *The Hitchhiker's Guide to the Galaxy* (New York: Del Rey, 1997).

14. Elon Musk, "Elon Musk's Full Speech at the 2017 International Astronautical Congress," Go To Space, September 29, 2017, video, 2:07, https://www.youtube.com/watch?v=cj3OgrvvBpE.

15. Bill Watterson, *It's a Magical World: A Calvin and Hobbes Collection* (Kansas City: Andrews McMeel Publishing, 1996).

Index

About the Authors

Photo by Michelle Fairless

Michael Ashley is a former Disney screenwriter and the author of more than 40 books on numerous subjects, including four bestsellers. He co-authored *Own the A.I. Revolution* (McGraw Hill) which launched at the United Nations and was named by Soundview as one of 2019's top business books. A columnist for *Forbes*, *Entrepreneur*, and *Becker's Hospital Review*, Michael is an in-demand speaker who keynotes for various professional groups, including Vistage, the world's largest executive coaching organization.

A former screenwriting professor at Chapman University, Michael's writing has been featured in KTLA, *Entertainment Weekly*, *HuffPost*, *Newsbase*, *The Federalist*, *Fast Company*, Accordant Philanthropy, *Coeur d'Alene Press*, *National Examiner*, the UN's *ITU News*, *The Orange County Business Journal*, *The California Business Journal*, *Chronicle of Philanthropy*, *The Columbia Missourian*, Fox Sports Radio, and *The Orange County Register*.

Photo © Sierra Space

Tom is the CEO of Sierra Space, a leading commercial space company. Prior to Sierra Space, Tom was the Chairman, CEO, and President of Aerion Corporation, a company developing supersonic civil aviation. Tom was the President of Northrop Grumman's Aerospace Systems sector, an $11 billion global advanced technology business with 23,000 incredibly talented team members. The team members were pioneers in space-based observatories, satellites, fully autonomous intelligent systems, combat aircraft, high-powered lasers, and microelectronics.

Prior to becoming the President of Northrop Grumman's Aerospace Systems sector, Tom was the President of their Technical Services sector, a recognized leader in integrated logistics and modernization, defense and government services, and training solutions, with more than 19,500 employees. Tom led businesses at more than 300 locations in all 50 states and 29 countries, supporting a diverse customer base, including the U.S. departments of Defense, Energy, Homeland Security, State, and the Interior, as well as NASA. Tom retired from Northrop Grumman in August 2017 after nearly 31 distinguished years of service with the company.

Tom is a successful operating executive with unquestionable integrity, broad industry experience, and proven strategic, financial, risk management, and technical skills. Tom is a visionary and inspiring leader, with relentless discipline and long-term focus.

Tom is genuinely respected and trusted by his customers, employees, industry peers, and investors. He has a very thoughtful, creative, and collaborative style. Tom has a comprehensive understanding of the government regulatory and compliance environment.

Tom has served on the board of the Smithsonian National Air and Space Museum, the board of governors for the USO, the board of trustees for the Florida Institute of Technology, Purdue University's School of Astronautics and Aeronautics, and Local Motors. Tom currently serves on the board of councilors for USC's Viterbi School of Engineering.

Tom is the recipient of the 2016 Daniel J. Epstein Engineering Management Award from the University of Southern California's Viterbi School of Engineering. The Engineering Management Award was founded in 1978 to recognize industry leaders who have made lasting contributions to science and engineering through exemplary professional accomplishments and contributions to the field of engineering management.

Tom earned a bachelor's degree in aerospace engineering from the University of Southern California. He has completed numerous advanced management programs including MIT's Corporate Strategy, Northwestern's Value Creation through M&A, Caltech's Management of Technology and Innovation, Caltech's Strategic Marketing of Technology Products, UCLA's Executive Marketing, and Defense Acquisition University's Advanced Program Management course.